SpringerBriefs in Public Health

T0254744

Series Editor
Macey Leigh Henderson

More information about this series at http://www.springer.com/series/10138

Travis N. Rieder

Toward a Small Family Ethic

How Overpopulation and Climate Change Are Affecting the Morality of Procreation

 Springer

Travis N. Rieder
Berman Institute of Bioethics
Johns Hopkins University
Baltimore, MD, USA

ISSN 2192-3698   ISSN 2192-3701   (electronic)
SpringerBriefs in Public Health
ISBN 978-3-319-33869-9   ISBN 978-3-319-33871-2   (eBook)
DOI 10.1007/978-3-319-33871-2

Library of Congress Control Number: 2016941751

Printed on acid-free paper

This Springer imprint is published by Springer Nature
The registered company is Springer International Publishing AG Switzerland

To Sinem,
who taught me the value of parenting:
Thank you for being our daughter.
I hope we're doing okay.

Preface

The idea that sparked this project began in another paper entitled, "Adoption, Procreation, and the Contours of Obligation," which was published in the *Journal of Applied Philosophy* in 2015. There, I investigated whether each prospective parent might have a duty to adopt rather than procreate, as there are several, powerful lines of reasoning in favor of such a view. Not only are there millions of adoptable children in need of the parenting resources (an argument made independently by philosophers Tina Rulli and Daniel Friedrich), but there seem to be moral concerns about the very creation of children in the first place. In particular, I raised concerns that each new person we create exacerbates global resource shortages and the threat of catastrophic climate change.

In 2014, I was discussing an early draft of this project with Macey Leigh Henderson, who was intrigued by the relationship between procreation and climate change. She asked if it was my view that overpopulation thus constituted a public health crisis. And, while I hadn't thought of it in exactly those terms before, I said "yes" and that the real challenge of my view was accounting for how such a crisis generates individual moral obligation. Macey responded that this sounded like an interesting addition to a new book series that she was editing on public health ethics, and the current project was born.

My thanks to the *Journal of Applied Philosophy* for publishing that original article, to my friends, colleagues, and critics, who have pushed me on its argument, and to Macey for extracting a more sustained treatment of the climate ethics portion of the argument. The bulk of the first draft of this manuscript was written during my time as a Hecht-Levi Postdoctoral Fellow in the Berman Institute of Bioethics at Johns Hopkins University, and I am grateful to the Hecht-Levi Program, as well as the faculty and my fellow postdocs, who provided an exceptionally congenial environment for doing bioethics. In particular, I want to thank Ted Bailey, Jeff Kahn, Stephanie Morain, Bridget Pratt, Kevin Riggs, Yashar Saghai, and Miriam Shapiro, all of whom provided valuable feedback on one or more chapter, and audiences at Georgetown University's Works-in-Progress Group, Southern Illinois University's Philosophy Colloquium Series, the Berman Institute's Faculty Workshop, and James Madison University's Philosophy Colloquium Series, for helpful dialogue and insightful advice. Marcus Hedahl deserves special mention for reviewing, in detail, the entire manuscript and providing invaluable criticism and suggestions, making the current draft far better than the one he saw. And finally, I want to thank my partner in crime and in life, Sadiye, whose thoughtful discussion of ethics and our own family is reflected in nearly every page written; indeed, it is difficult to discern where her influence ends and my own ideas begin.

Baltimore, MD, USA   Travis N. Rieder

Contents

Chapter 1
Global Population and Public Health

It took all of human history up until the 1800s for the global population to reach one billion. The most recent billion, however, was added considerably faster. As in: the global population grew from six to seven billion in approximately 12 years. Our population is growing so fast that anyone alive today who was born prior to the mid-1960s has seen the population *double*.[1] In other words: we humans have made a lot of people very quickly. The concern that will occupy me in the rest of this short book is that we now have very good evidence that we made too many.

What does such a claim—that we have made too many people—mean? It is not as if every nook and cranny of the earth has a human on it. It's not that the earth is literally filled to the top with people.[2] However, people need resources, and some people use more of these resources than others. So while I am not concerned at all about how many people the earth can *hold*, I am quite worried that the earth cannot *sustain* any more people—indeed, that it likely cannot sustain us. In this chapter, I will provide the evidence for this claim, citing a wide variety of problems that our numbers either give rise to or make worse, but with a special focus on what is arguably the direst of these worries: climate change. I will conclude that there is a global population crisis, and that this crisis constitutes a public health emergency. The fact of this emergency, then, raises the question: is there something that *I* should do about it? Do the intersecting problems of population and climate change generate a private moral burden for me, for you, and for everyone else?

---

[1] Population data is easily located in any number of places. For some of the best, up-to-date information, see the UN's Population Division, found at http://www.un.org/en/development/desa/population/index.shtml (last accessed January 27, 2016).

[2] This point about the ambiguity in the discussion of earth's carrying capacity was made strongly by Joel Cohen in his (1996).

© The Author(s) 2016
T.N. Rieder, *Toward a Small Family Ethic*, SpringerBriefs in Public Health,
DOI 10.1007/978-3-319-33871-2_1

1.1   How Many People Can the Earth Sustain?

It's unclear exactly what it means to ask how many people the earth can sustain. On the one hand, this seems like an empirical question: we will need to turn to scientists to tell us about resource use and availability, and to demographers to tell us about populations and their behaviors. However, this initial inclination only gets things partially correct. The empirical facts matter—*a lot*—but the question is also one of values and morals. To see this, let's consider a few different scenarios.

A major worry raised about the global population is that, although population growth *is* slowing down, it is not doing so quickly enough. We are on track to be at nine or ten billion by 2050,[3] and so a question that gets asked a lot is whether the earth can sustain a population of ten billion people. However, the answer to that question is, in one sense, unequivocally 'yes'. If those ten billion people renounce all unnecessary greenhouse gas-producing activities, turn to a sustainable vegetarian diet, and live simple lives, then there is no reason to think that the world cannot support a population of ten billion. Call this fictional version of our future **Modest World**.

On the other hand, we might think both that such a conversion by the world's wealthy is unlikely, and that we have a duty of justice to pull some of the world's poorest people out of poverty, increasing their resource consumption. Can the earth support a population of ten billion people, some of whom are fantastically well-off, and the rest of whom are living decent lives? Let's call this case **Excess World**.

Finally, we can even consider what is likely to be the *actual* constitution of a population of ten billion people: a population much like ours, only bigger. Such a population has some fantastically wealthy people, who consume a vast majority of the planet's resources, and then very, very many poorer people, who live modest or desperate lives, and who consume far fewer resources. Perhaps this is the population that, as a matter of realism, we ought to be most concerned with, so let's call this one **Real World**. Can the earth sustain this version of our future selves?

The answers to these questions are illuminating. It is exceedingly unlikely that the earth could support **Excess World** or **Real World**. In fact, the data suggests that the earth cannot sustain our *current population*. According to the Global Footprint Network we are already in an ecological 'overshoot', which means that the global population is using resources faster than they can be replenished. This means that the earth will, at some point, be unable to provide for our population, even without

---

[3] According to the UN World Population Prospects 2015, we should expect the global population to hit 9.7 billion by 2050, and 11.2 billion by 2100, based on their 'medium variant' fertility projections. However, the medium variant fertility projection has been consistently low for several years, prompting the UN to repeatedly update their projections. We might then worry that the actual future will look more like the higher limit of the confidence band (the highest prediction within a 95 % confidence interval), which predict a population of more than 11 billion by 2050, and a population of 13.3 (!) billion in 2100 (United Nations, Department of Economic and Social Affairs, Population Division, 2015).

more growth. In essence, we are charging our ecological credit cards for more than we can cover, and at some point, the earth's bank will simply refuse to extend us any more credit.[4]

Exactly how much are we overshooting our carrying capacity? The GFN estimates that we currently would need about 1.6 earths to cover our consumption, which means that each year, we overcharge our ecological credit card by about 60 %. However, that takes the global population as it is—radically unjust, with billions of people living in poverty. So we might wonder how many people the earth could sustain if everyone lived good lives—the kind of lives we might want them to live. This is a tricky question, and there is no agreed-upon answer. However, an important study by a group of scientists in the late 1990s estimated that the earth could support a global population of only about two billion, if everyone consumed as Western Europeans did (Pimentel et al., 1999). More than a decade later, environmental scientist Tim De Chant utilized data from the Global Footprint Network to estimate that, if everyone consumed resources as Americans do, we would need about 4.1 earths to sustain a population of seven billion (http://persquaremile. com/2012/08/08/if-the-worlds-population-lived-like/).

Clearly, then, the earth cannot sustain **Excess World** or **Real World**—barring an unprecedented advancement in various technologies—as the earth cannot even sustain *us*. So the earth could possibly hold ten billion modest vegetarians, but could not sustain ten or even our current 7.3 billion people when some are as wealthy as we are. Further, the earth can sustain a shockingly small number of people if everyone were as wealthy as the average member of the developed world. And so now we can see why the question, 'How many people can the earth sustain?' is partially a value question, as the number of people the earth can sustain depends at least partially on the goodness of those peoples' lives and the justness of the distribution of wealth.

The earth, then, cannot sustain the populations in **Excess World, Real World**, or even our current, actual world. What, exactly, does that mean, though? Sustaining a population isn't a matter of, say, having a patch of land for each person to stand on, so what are the *problems* raised by our numbers? Let us now turn to exactly this question.

1.2 The Multiplier Effect: Food, Water, Energy and Climate

It is said that population is the 'multiplier of everything', as any problem caused by humanity is made worse by our growing numbers (Ryerson, 2010). If one person doing something is a problem, then very many people doing it is a big problem. Unfortunately, today we are faced with a great many problems that our growing population multiply. I will here briefly survey a few.

---

[4] See this, and related data, at the GFN's website: http://www.footprintnetwork.org/en/index.php/ GFN/page/world_footprint/ (last accessed January 27, 2016).

The most obvious problem with a large population is that each person needs to eat and drink, and food and water are limited resources. In fact, we already have scarcity of each in many parts of the world.[5] Further, people consume energy resources in order to survive and in order to flourish in various ways; but our primary sources of energy—fossil fuels and coal—are running out, and doing massive damage to the natural environment in the meantime (as we will discuss in more detail, below).[6]

The most important point, here, is an utterly obvious one, but one which clearly needs stated: the earth's resources are finite, and finite resources cannot support an infinite population. That means that there is some finite number of people, using resources according to some distribution, that the earth cannot sustain. The models from the previous section suggest that we have passed that number and distribution.

Of course, population is not the only variable at play. Our resource use, technology, waste, and other behaviors all matter as well. And so one might object that the problem is not necessarily the number of people, but our failure to develop and implement the appropriate technology, or to change our behaviors. While it is true that there are other controllable variables, we might be too late to rely on them. This is especially true given the final problem that I will discuss, which is global climate change.

Although the earth's population raises or multiplies many problems, in this book, I will focus particularly on climate change, for a variety of reasons. First, it is a major cause of many other, population-sensitive worries. The warming of the planet and the disruptions brought about by such warming will cause and amplify all of the challenges mentioned above—food, water, and energy. As our climate changes, we can expect to experience more food shortages, severe water stress, and the need to find alternative energy sources as we attempt—too late—to slow the cooking of the earth. Thus, even if there would otherwise be enough of these resources, climate change will result in hardship in each of these arenas.

Additionally, climate change will kill and otherwise disrupt the lives of billions of people in the future, *even if we don't run out of other resources*. Rising sea levels, desertification, extreme heat waves, changes in disease vectors, and more frequent

---

[5] For a visual snapshot of the world's food and water shortages, see the FAO Hunger Map 2013 (The Statistics Division, Food and Agricultural Organization of the United Nations, 2013); for more details, explore the UN's full website at www.fao.org/economic/ess (last accessed January 27, 2016).

[6] This problem is really two problems, as it *is* likely that our resources will run out if we let them (if we continue use, unabated), but letting our resource use get to that point would be an environmental disaster. Even as I finish this manuscript in early 2016, America continues to look for novel (and ever dirtier) sources of fossil fuel, from the Alberta tar sands, to ever more wide-spread use of fracking, to the opening of drilling in the (now melted from climate change) arctic. And what all of this ignores is that, even burning through what is already on the accounting books for major energy companies will guarantee absolutely catastrophic climate change. In fact, according to Bill Mckibbin, the amount of coal, oil and gas in reserve (what we are planning to burn) is *five times* the amount that would lock in 'dangerous climate change' (Mckibben, 2012). So on the one hand, there is in fact a hard limit to the amount of fossil fuels that we can extract from the earth; but on the other hand, we will run out of what we can *safely* burn, long before we run out of what we can *possibly* burn.

and more extreme weather events will result in death, destruction, and challenges to global political and economic systems. Climate change has the potential to be the biggest moral tragedy in the history of humankind.

So how does population relate to climate change, exactly? Well, in a rather unsurprising way. Climate change is caused by the accumulation of greenhouse gases (GHG) in the atmosphere, and the emission of these gases is a product of not only the activity of individuals, but of the *number* of individuals doing the emitting. Precisely as one would expect, then, data from the US and Europe show that GHG emissions rise with the population in a near perfect 1:1 correlation (Ryerson, 2010, p. 3). This is why it is common to hear climate change arguments of the following form: we are producing too much GHG; further, by 2050 our population will have grown to nine or ten billion; thus, to reduce our emissions *while increasing our population*, we must act quickly and decisively to curb our *per capita* output. That is: because our population is growing, we must change our emitting practices *a lot*.

Although the argument above is accurate—if we expect to slow climate change while growing our population, we must act quickly and decisively—there is obviously another lesson that we could take from the data. We *could* note that there are two variables at play: the emissions of each individual (per capita emissions) and the *number of individuals*. From this observation, we could then draw the conclusion that our solution to climate change must address either or both variables. Over the course of this book, I will essentially be making the case that, given the multiplier role of population in the climate change problem (as well as others), we should take very seriously the idea that the dire moral threats of the day give us reason to address the population variable as well as the emissions variable (and further, that the need to address the population variable generates a moral burden on each individual to consider changing her procreative habits).

1.3   The Role of Population in Mitigating Climate Change

One reaction to the move suggested above—that of taking population seriously as a variable in the climate change equation—is that changing procreative behaviors is an intensely *personal* sort of intervention, and so we should avoid it if we could. Thus, since there is another variable that we can work on—namely, *per capita* GHG emissions—we should do that.

This objection is understandable, and if the global population were showing any signs of working hard enough to mitigate the dangers of climate change, I would take it to be correct. However, we aren't. As we will discuss more in the next section, climate change is already occurring, and people are already suffering; further, worse changes are on the way, and we are not on anything like a course to prevent them. Before talking about these dark predictions, however, let's look briefly at some calculations concerning the effects of focusing on emissions vs. the effects of focusing on reducing fertility rates.

In 2008, a biologist named Frederick Meyerson calculated that if we wanted to hold steady the total global emissions at (then) current levels, while keeping up with population growth, we would need to reduce global, average, *per capita* emissions by 1.2 % every year. Now, there are a couple of things to note about this number. First is that this is what would be required *just to maintain current levels of emissions*, which won't be good enough; we ultimately need to *reduce* our total emissions. Second is that, although a 1.2 % reduction might seem realistic, it is not. According to Meyerson's calculations, we have not been able to reduce *per capita* emissions by even 1 % *over the course of the previous 38 years* (Meyerson, 2008). Thus, to believe that we can act so as to mitigate climate change by changing only our carbon-emitting behaviors requires believing that we can do more to reduce emissions every year than we have been able to accomplish in the past four decades. This seems unlikely.

In addition, the Intergovernmental Panel on Climate Change (the IPCC) has surveyed over 900 scenarios for mitigating climate change, and has recommended a method for avoiding dangerous climate change. Unfortunately for us, what they found is the following: in order for us to have a 66 % chance or better of avoiding a 2 °C rise in global average temperature over preindustrial times, we must make radical, decisive movement towards a decarbonized economy *now*. In fact, change in behavior by itself is likely insufficient: if we are to avoid such dangerous climate change, we must successfully implement as-yet-unproven technologies such as 'Carbon Capture and Storage' (CCS), which would take existing carbon out of the atmosphere and bury it underground. We would likely also need to ramp up production of nuclear energy, which has its own dangers and environmental costs (Intergovernmental Panel on Climate Change, 2014, p. 10). In short: the most thorough study to date concerning our chances of mitigating climate change through behavioral change requires: radical, unprecedented action by the global community; the existence of a technology that has not yet been proven; and an increase in other, potentially dangerous technologies such as nuclear power.

What data like this seems to indicate is that focusing our climate change mitigation efforts solely on carbon-emitting activities is not sufficient. So long as the population keeps growing at its predicted pace, it will take more effort than the global community has been willing to provide, as well as massive technological effort, in order to really curb our GHG emissions. But is there really any reason to think that changes in the global population would be more effective?

In fact, there is. As we will discuss in more detail later, creating a new person is among the most carbon-intensive activities that most people ever engage in. As a result, fairly modest changes to the population can have impressive results. For example, a group of scientists recently asked what might be accomplished for the environment under alternative fertility scenarios. Their findings? If the global fertility rate were reduced by a modest amount—an amount that would be possible completely without coercion, through simply providing health care, education, and family-planning services to the poorest people in the world—the annual global savings would amount to *5.1 billion tons of carbon by the year 2100* (O'Neill et al., 2010). To put that number in perspective: in 2013, total carbon emissions amounted to 9.6 billion tons. A modest reduction in fertility, then, could amount to a yearly savings by 2100 of more than half our current yearly emissions.

Global population, then, is a very real variable in the climate change equation. If we could change the number of people on earth, this would have a profound effect on our ability to combat climate change. Further, if we can't change the number of people on earth, we may have no real hope of reducing overall GHG emissions quickly enough to save millions of people from the effects of climate change. Our growing population, then—in addition to whatever other problems it raises concerning scarce resources—is of serious concern as it relates to climate change.

1.4 Moral Urgency

The case for treating the global population as an important variable in the climate change equation looks powerful. Slowing and eventually reversing our total GHG emissions will be very difficult if we must hold population growth steady, while even modest adjustments to the global fertility rate can have dramatic effects on our total emissions. But even so, I have not yet established that there is a population *crisis*. For all I've said thus far, it may be that climate change isn't all that bad (or at least, isn't so much worse than making sacrifices that would avoid it), or that we have plenty of time to mitigate its harms. Unfortunately, neither of these claims is true. The harms of climate change promise to be immense, and they are coming much sooner than most people realize.

It is widely claimed, as in the IPCC report, that 'dangerous' climate change corresponds to an approximately 2 °C global average temperature rise over preindustrial times. Such an increase is expected to occur as a result of atmospheric carbon dioxide (CO_2) reaching approximately 450 ppm. So what will happen when we reach this point? The IPCC predicts a sea level rise of several feet, as water expands according to its temperature and the world's glaciers melt (Intergovernmental Panel on Climate Change, 2014). The warming water will disrupt climate patterns and increase the frequency and severity of storms, while the rising sea-level makes the effects of these storms even worse for coastal and low-lying areas. Flooding and storm surge will be a constant battle. Bangladesh, which is among the most at-risk of all nations to various climate change harms, will experience near-constant flooding in the delta regions, along with catastrophic water stress and food shortages (Intergovernmental Panel on Climate Change, 2014, Chap. 24: Asia). Low-lying coastal nations such as the Maldives, Tuvalu, and Kiribati will be forced to systematically evacuate residents from their lowest points, before finally abandoning their home nations to the rising tide altogether.

Rising seas will not only affect those in developing and island nations. Many European countries, such as England and the Netherlands, are already evaluating adaptation strategies so that they might withstand at least the initial assaults from the water (Intergovernmental Panel on Climate Change, 2014, Chap. 23: Europe). Parts of the United States, however, have failed to begin investigating adaptation possibilities, and may suffer catastrophic loss as a result. Miami Dade County, for instance—home to the famous Miami Beach—sits on a bed of porous limestone rock, and so is particularly susceptible to rising tides and storm surge. For this reason, it has been estimated that as little as a one foot rise in sea level could spell disaster for the county, contaminating the fresh water supply, backing up the sewer system, while flooding extensively during each of the (ever more common) storms that hit the area (McKie, 2014). If accurate, such a prediction could spell disaster for Miami within the next few decades.

Of course, sea-level rise is just one aspect of the coming climate changes. In other areas, we will see increased desertification and more deadly droughts and heat waves. Water stress will become more common, as will food shortages, while the global economic system attempts to adapt to the massive changes in growing seasons and crop yields. Over the coming years, *millions* of people will die, and many millions more will be dislocated—the poorest of these, with nowhere to go, will become climate refugees. In short: the effects of climate change—if we do nothing to stop its arrival—will be catastrophic.

How soon are such effects coming? Disconcertingly soon. In 2014, for the first time in *at least 800,000 years*—and likely the first time since the Pliocene era, between 3 and 5 million years ago—atmospheric carbon climbed to 400 ppm. In light of this, climate scientist Michael Mann—lead author of one of the most important papers in the climate change literature (Mann, Bradley, & Hughes, 1998)—revisited his calculations concerning the future. And according to his new predictions, we will see a 2 °C rise by the year 2036 (Mann, 2014). That is very bad news, of course, as that is the point at which many of the worst climate disruptions will really get going. However, it is slightly misleading to say that 'dangerous' climate change will begin at that point. In fact, climate change is occurring *now*, and how dangerous it is depends on where you live. According to the IPCC, we are already seeing the sea level rise and the resultant increase in the severity and rate of flooding. In addition, we are now seeing, all over the world, an increase in extreme weather events of all kinds, more frequent and more extreme droughts and wildfires, expanded range of water-borne illnesses and disease vectors, biodiversity loss, and crop yield decreases (Intergovernmental Panel on Climate Change, 2014). In the summer of 2015, as I was working on this manuscript, an unprecedented 'heat dome' settled over the Middle East, causing deadly heat-index figures of 165 °F. Temperatures were so high, in fact, that the Iranian government was forced to announce a 4 day federal holiday, in order to protect people from needing to go outdoors. At the same time, the western US was experiencing catastrophic drought, which contributed to devastating forest fires. This sort of news is becoming less and less of a surprise, and we should expect things to get steadily worse.

In short: we are running out of time to mitigate the worst effects of climate change, and every day that passes is another day that we make the problem worse rather than better. When superstorms like Katrina and Sandy make landfall, we have to ask ourselves: did we do that? Would that storm have happened if we had taken more action against climate change sooner? The time-sensitivity of the problem, combined with the catastrophic costs, makes the climate change problem *morally urgent.*

1.5   Conclusion: The Population Crisis is a Public Health Emergency

The main lessons of this first chapter are (1) that population is a major driver of climate change, in addition to raising concerns about other limited resources; and (2) that climate change is a morally urgent problem. As a result, it seems appropriate to say that we have a *population crisis*—that the size of our population generates a problem that is massive in scale and dire in consequence.

The final observation that I want to make here, then, is that the population crisis presents us with a particular kind of threat—namely, one in 'public health'. A failure to mitigate climate change is a failure to adequately protect the well-being of the population as a whole, albeit while allowing disproportionate harm to the poor and the weak. But who, exactly, fails the population? Who is responsible for the harms of climate change? It is difficult to say, but whatever the answer is, it seems that the relevant moral agent must be some group or groups. The global community perhaps? Individual nations? The wealthy?

These observations reveal the difference between discussing morality in the context of the population crisis and the morality of, say, an individual killing. Unlike the case of murder, in which we can clearly identify an individual who is responsible for the killing, and who is therefore wrong, the population crisis is a moral problem for an *aggregate*; it is a problem for, well, *populations*. But this leads us to a particular challenge, as public health emergencies—such as the 2014 Ebola outbreak in West Africa, for example—are typically handled by governments, regulations, and policy interventions. *I* cannot stop an outbreak, but a coalition of the world's governments can. Similarly: *you* cannot reduce driving deaths by a large percentage, but a public health intervention in the form of a seat-belt law can.

In these cases of public health problems and emergencies, is there a moral burden on individuals? Must I, for instance, donate money to Doctors Without Borders to help them in the fight against Ebola? *This* is the form of question that will occupy us for the rest of this book. Although no one of us can solve the population crisis, we all make decisions relevant to making the problem better or worse—that is, we all make procreative decisions. Must I, then, refrain from procreating? Or should I at least refrain from creating *too many* people? What kind of responsibility is it plausible to say that I individually inherit as a result of a public health emergency? Is it possible that I have a *duty* or *obligation* not to procreate?

Let us call the class of potential duties that would require us to have either no children or few children *procreation-limiting duties*. For those who find it plausible that large, aggregate moral issues such as public health emergencies generate individual obligation, it should seem disconcertingly plausible that each of us has procreation-limiting duties. Overpopulation constitutes a massive public health crisis, contributing dramatically to climate change in addition to other resource shortages. If the Ebola outbreak in West Africa gives each of us an obligation to donate to Doctors Without Borders, or if tragedies such as devastating tsunamis in Southeast Asia can obligate us to donate time, money or resources to organizations like Oxfam, then it looks like we may inherit the obligation to do our part in slowing population growth.

The goal of this book is to investigate, rigorously and systematically, precisely that hypothesis—that each of us might inherit procreation-limiting duties as a means to combat overpopulation and climate change. And my conclusion (sorry to give it away so early!) will be that something disconcertingly close to this suggestion is true. We have, I think, something that we can call a 'moral burden' concerning our procreative choices, and this leads me to what I call a *small family ethic*.

Such a view is not, however, popular; indeed, I would prefer that it not be true. I have a child myself, and think that the project of creating and rising a child can be among the most meaningful in human experience. But that project comes with costs, and those costs are largely born by others—most of whom are less fortunate than you and I are. These considerations lead me to think that even this, most intimate of decisions, is subject to a demand for justification.

We start, then, with the conclusion of the present chapter: that there are too many people on earth, together emitting far too much GHG much too quickly. And that the public health crisis of overpopulation leads to the intuitive conclusion that morality might demand of each of us that we not contribute to such a crisis. In other words: the very facts of the matter seem to suggest that each of us is subject to procreation-limiting obligations. And the question for the rest of the book is: could that really be true? Might it really be the case that morality requires that we limit the size of our families?

In attempting to answer this question, I will be as sympathetic as possible to what I am sure will be a vocal opposition. It is hard to believe that we could have such a burden, and so I have structured what follows as a steady stream of challenges to the view that we do. I will even concede many points along the way, weakening the supposition of what morality may require; indeed, I will occasionally do this even when I don't believe the concession I am making, if the argument in favor of moral requirements seems too uncertain. I adopt this strategy, because I want to know what our moral burden may be, *even if the most restrictive arguments fail.* I want to know the answer to the question: if I can show that we don't have a strict obligation not to have children, are we therefore off the moral hook?

I begin in the following chapter, then, with a powerful objection to the idea that we can have an individual obligation to fight such a massive crisis. In short, the objection claims that if my having a child won't make any meaningful difference to the amount of harm caused by climate change, then morality can't require that I not have a child on account of the prospect of harmful climate change. Many people find such reasoning plausible, but I argue that it fails to account for an entire class of moral obligations that we do tend to think we have. After all, my throwing a paper cup out the car window makes no real difference to how much anyone is harmed by problems of waste management; but we tend to agree, I think, that we each have a duty not to litter.

If we do sometimes have a duty not to act in ways that don't really make a moral difference, then there must be some explanation of that fact. In Chap. 3, then, I propose three candidate moral principles that could explain the wrongness of acting in ways that don't make a difference to what seems to be a moral problem. Although it's not perfectly clear what one's procreative obligations would be in light of these principles, each one would justify *some sort* of procreation-limiting obligation. And each principle, I argue, is overwhelmingly plausible as a candidate moral duty. Is that, then, the end of the story?

Not quite, it turns out. Even if the candidate principles of Chap. 3 are valid, it is a separate question as to whether they successfully entail any particular moral obligation. Many considerations can get in the way of a valid moral principle justifying a particular obligation, and one very relevant such consideration is that of *demandingness*. A duty not to procreate (or even a duty to limit one's procreative behaviors) might be thought to be overly *demanding* in a way that undermines its plausibility. In particular, one might argue that an obligation that 'robs one of her integrity as a moral agent' is demanding in a way that undermines its validity. Further, one might think that having certain moral rights—such as robust *procreative rights*—would block the application of procreation-limiting duties. I investigate all of these options in Chap. 4.

It is very difficult to determine just how far these objections go. I suggest that they quite plausibly undermine a proposed obligation to remain childless forever, and so it is very likely not a violation of duty to have a single child. The question gets much more difficult when considering having more than one child, though, and I do not claim to determine whether there may be *any* valid procreation-limiting duties (a duty, say, to have no more than one or two children).

Unfortunately for those of us who might wish to be off the moral hook regarding our procreative behaviors, the success of Chap. 4 is limited. Even if having a child, or multiple children, is within one's rights, that does not make it the *right thing to do*. Indeed, as I argue in the book's closing chapter, morality can tell us much more than merely what our rights and duties are. Morality can also tell us what is *recommended*, what is *blameworthy*, and what is *virtuous or vicious*. Rights and duty matter, but so do these other things. A full, rich picture of the moral landscape will include a variety of moral considerations, all of which can affect our evaluation of procreative behavior. What the arguments of this book suggest to me, then, is that even if we have fairly robust procreative rights, and so are not subject to procreation-limiting duties, morality may yet judge us harshly for unrestrained procreative behavior. Full consideration of the relevant principles, reasons, virtues and attitudes lead me to support a small family ethic.

References

Cohen, J. E. (1996). *How many people can the earth support?* New York: W.W. Norton.
Intergovernmental Panel on Climate Change. (2014a). *Climate change 2014: Impacts, adaptation, and vulnerability.* Cambridge: Cambridge University Press.
Intergovernmental Panel on Climate Change. (2014b). *Climate change 2014: Mitigation of climate change.* Cambridge: Cambridge University Press.
Mann, M. (2014, April 1). Earth will cross the climate danger threshold by 2036. *Scientific American.* Retrieved from http://www.scientificamerican.com/article/earth-will-cross-the-climate-danger-threshold-by-2036/.

Mann, M. E., Bradley, R. S., & Hughes, M. K. (1998). Global-scale temperature patterns and climate forcing over the past six centuries. *Nature, 392*, 779–787.

Mckibben, B. (2012). *Global warming's terrifying new math*. Retrieved from Rolling Stone http://www.rollingstone.com/politics/news/global-warmings-terrifying-new-math-20120719?page=2.

McKie, R. (2014, July 11). Miami, the Great World City, is drowning while the powers that be look away. *The Guardian*. Retrieved from http://www.theguardian.com/world/2014/jul/11/miami-drowning-climate-change-deniers-sea-levels-rising.

Meyerson, F. B. (2008, January 17). *Population growth is easier to manage than per-capita emissions*. Retrieved from Population and Climate roundtable discussion held by the Bulletin of the AtomicScientistshttp://thebulletin.org/population-and-climate-change/population-growth-easier-manage-capita-emissions.

O'Neill, B. C., Dalton, M., Fuchs, R., Jiang, L., Pachauri, S., & Zigova, K. (2010). Global demographic trends and future carbon emissions. *Proceedings of the National Academy of Sciences, 107*, 17521–17526.

Pimentel, D., Bailey, O., Kim, P., Mullaney, E., Calabrese, J., Walman, L., et al. (1999). Will limits of the earth's resources control Human numbers? *Environment, Development and Sustainability, 1*, 19–39.

Ryerson, W. (2010). Population: The multiplier of everything else. In R. Heinberg & D. Lerch (Eds.), *The post carbon reader: Managing the 21st century's sustainability crises*. Healdsburg, CA: Watershed Media.

The Statistics Division, Food and Agricultural Organization of the United Nations. (2013). *FAO Hunger Map 2013*. Retrieved from Food and Agricultural Organization http://www.fao.org/fileadmin/templates/hunger_portal/docs/poster_web_001_WFS.pdf.

United Nations, Department of Economic and Social Affairs, Population Division. (2015). *World Fertility Patterns 2015—Data Booklet (ST/ESA/SER.A/370)*.

Chapter 2
What Can *I* Do? Small Effects and the Collective Action Worry

Large-scale problems like the population crisis can leave each of us, as an individual, feeling causally impotent in our ability to make a difference. While it is technically true that I can make the population crisis better or worse—that is, most of us can choose either to make more people or not—in the context of a population of more than seven billion, the number of people any one of us can create appears not to matter. Consider resource shortages: it seems absurd to think that my adding one child, or two, or even seven, will make any real difference. If there is not enough food, clean water, energy resources, or carbon sinks to provide for the population, that will be true whether I procreate or not. The idea that we might have sufficient resources for 7,300,000,000 people, but not enough for 7,300,000,001 likely strikes us as absurd. But if my having a child doesn't change whether or not there are enough resources, and no individual will perceive the difference, then it would appear that my action doesn't actually harm anyone. The existence of another person in the world is a difference that doesn't make a moral difference. This sense of causal impotence arises from the *scale* of the problem.

2.1 The Scale of the Problem

In the context of climate change, the worry about causal impotence is frequently raised. After all, the limited resource here is the atmosphere's ability to accept some amount of GHG without it violently disrupting the climate, and the scale of the problem is virtually unimaginable. For instance: the average American will emit, in her lifetime, approximately 1644 metric tons of CO_2 (Murtaugh & Schlax, 2009, p. 18). How does that compare to the total emissions contributing to the problem of climate change? Every year, we emit a total of more than *30 billion* tons of CO_2.[1] So

---

[1] It is important to note that both scientists and governments regularly use two different measurements to represent climate-affecting GHG: carbon dioxide, and carbon. This can lead to confusion,

© The Author(s) 2016
T.N. Rieder, *Toward a Small Family Ethic*, SpringerBriefs in Public Health,
DOI 10.1007/978-3-319-33871-2_2

the average lifetime total of even an American—who has some of the highest emissions in the world—is a vanishingly small fraction of the global population's yearly contribution.

The scale of the problem leads to a collective action worry, which is to say that the harms of the population crisis will occur only if a sufficient number of people procreate, and that those same harms can be avoided only if a sufficient number of people refrain from procreating. What makes this really difficult as a problem, then, is that so long as there is no reason to expect collective action towards solving the problem, it appears that there may be *no reason* for me not to procreate, as my procreating makes *no significant difference*. It makes a technical difference, sure (it adds a single individual to the population); but that single individual doesn't cause the crisis, and refraining from adding that individual does not contribute meaningfully towards solving the crisis. This seeming fact of being unable either to cause or to significantly mitigate a large-scale crisis is what I earlier called 'causal impotence'.

It is likely clear that the worry is not specific to procreation in the context of climate change—the collective action worry is a problem for climate change ethics in general; in fact, Dale Jamieson calls climate change "the world's largest and most complex collective action problem" (Jamieson, 2014, p. 162). Thus, if there is a question as to whether procreating is morally problematic in the context of climate change, this could be either because procreating is unique in its ability to cause the harms of climate change, or because there is a more general response to the Collective Action Worry. In what follows, I will consider both of these strategies.

According to the first argument I consider, one might think that procreating makes more of a difference to the problem of climate change than you might think—especially for those of us who are wealthy by global standards. This is because procreating has a *carbon legacy*, which is to say that it is the GHG anti-gift that keeps on giving. This argument would work if the fact of carbon legacy were so severe that procreating did, in fact, seem to make a significant difference to the problem of climate change.

Although the fact of carbon legacy is crucial to understand, I will suggest that the first strategy is unlikely to work. While procreating causes a significant amount of GHG *relative* to other actions an individual can take, it does not cause significant emissions *all things considered*. We are not off the moral hook yet, however, as I will suggest that our common intuitions about the environment suggest a second argument, which is that we may regularly be obligated not to contribute to harms, *even if our contributions aren't significant*. This section will involve a bit of moral theory, as the primary goal is to suggest that it is a mistake to think that *making a*

---

as carbon dioxide is obviously heavier than carbon. As a result, one might find herself in the situation that I find myself here, having cited in the previous chapter a global, annual emission of 9.6 billion tons of carbon, and then citing here a global, annual emission of more than 30 billion tons of CO_2. There is no inconsistency here, as CO_2 is approximately 3.67 times heavier than carbon, and so the math works out. However, it is inconvenient (and potentially confusing) that the relevant studies in the previous chapter employ carbon measurements whereas the studies cited here employ CO_2 measurements, and so the reader must be alert to the measurement units.

significant difference is the only possible justification for a requirement not to contribute to a massive harm; indeed, I will suggest that many of our widely-shared
moral judgments rely on the idea that we are sometimes obligated to act or refrain
from acting, regardless of whether our contribution to a problem is *significant*. Now,
this suggestion will write a check that I don't cash until the following chapter, as
these judgments require some account of what might make them true, and I don't
provide such an account until Chap. 3. However, I take it to be important to motivate
the intuition first.

Finally, it's worth forecasting now that the following discussion of significance
will come back as relevant in the discussion of candidate moral principles in Chap.
3. This is because if we are sometimes obligated to refrain from an act that doesn't
significantly cause the likely harms of everyone's performing that act, the *relative*
contribution that one's act makes to those harms may yet strengthen or weaken the
obligation. That is: if I am obligated to minimize my carbon footprint (despite the
collective action worry), then I have a stronger obligation not to fly across the
Atlantic for a weekend than to unplug my television when I'm not using it. And if
this is the case, then the fact of carbon legacy comes back as important yet again,
since procreating is likely the most carbon-intensive activity that most of us could
engage in. As a result, I will suggest that if there are *any* individual moral obligations due to climate change, these would likely include procreation-limiting
obligations.

2.2 Lessons from Climate Ethics

All of us (at least, anyone who might be reading this) engage in unnecessary carbon-
emitting activities. We drive for pleasure, take vacations, use electricity to watch
television, and a million other things. My own preferred vice is riding a motorcycle
for pleasure — not to get anywhere, but because being on a motorcycle is fun. But of
course, most motorcycles (including mine) use internal combustion engines, and so
require fossil fuels, which it then burns for energy, resulting in CO_2 and water as
waste. And the most fun way to ride a motorcycle is fast, at a race track, where fuel
consumption is especially high. So here is a fact about me: purely for fun, I have
chosen to occasionally take my motorcycle to the track and burn several gallons of
fuel. While doing this has been a great joy in my life, my life is full of other great
joys, and I could be happy without it. So now we have our question: is it morally
permissible for me to take my bike to the track? Given that doing so is just for fun,
and that my life would be good without it, can I justify the resource use of this activity, and my resultant small contribution to climate change?

This sort of question has occupied many ethicists. Walter Sinnot-Armstrong, for
instance, argues that, as a result of the scale of the problem discussed above, there
are no moral theoretic justifications for claiming that my doing so is impermissible
(Sinnot-Armstrong, 2010). After all, my contribution to climate change through this
activity is miniscule, and so we face the same catastrophic problem regardless of

whether I take my bike to the track. An argument of this form points out that what makes climate change bad is the harms that it causes, but that individual actions play virtually no causal role in producing these harms. As a result, it looks like I am under no obligation to refrain from individual activities that contribute to climate change. As I admitted in the previous section, this sort of argument is intuitively compelling.

However, I am unsure that such arguments, which gain so much traction from an emphasis on causal impotence, are ultimately sound. The problem is that they assume some principle like the following:

Significant Difference: If the consequences of an act make no significant difference to the extent or severity of a moral problem, then the agent is not morally required to refrain from acting in light of the moral problem.

This language of making a 'significant difference', or perhaps a 'meaningful' or 'real' difference, is intentionally vague, and is supposed to highlight the fact that making a mere technical difference to some problem doesn't always matter. So it might sound natural to say that, although I can determine whether my trash goes into landfill or recycling (a technical difference), it doesn't make a *real* or *significant* difference, due to the scale of the problem. I could go around throwing my trash wherever I wanted for my entire life, and by itself, my activity wouldn't have a meaningful effect on the problem of waste management. As mentioned above, this sort of problem is one concerning *collective action*, as it is only when my activity is joined with similar activities of billions of other people that it becomes a serious problem. The principle of **Significant Difference**, then, can be said to capture a moral feature of Collective Action Problems.

2.3 The Carbon Legacy of Procreation

If **Significant Difference** were true, then the only way in which climate change could imply procreative obligations for individuals would be if the act of procreating made a significant difference to the problem of climate change. Since the problem of climate change concerns the harm that will be caused by climate disruptions, the relevant difference concerns this harm: one could be obligated not to procreate if procreating made a significant difference to the amount of harm resulting from climate change. Now, the general belief in climate change ethics is that *no* individual activity makes a significant difference to climate change, and so we should expect that this move is a non-starter. However, it's important to pause here and point out that procreating is different from any other single activity in which the average person engages, and it's different in a way that makes its emission effects truly massive. While I will not, ultimately, argue that procreating makes a 'significant difference' to the problem of climate change, I will suggest that consideration of procreating both reveals an ambiguity in the principle of **Significant Difference**, and helps to set up what is ultimately a successful refutation of that principle.

In order to see how procreating is unique, let's return to a more mundane case of emitting activity—that of my taking my motorcycle to the track. If **Significant Difference** is plausible anywhere, it is likely plausible here. When I engage in this activity, I get a lot of pleasure out of an activity that takes less than 10 gal of fossil fuel. Further, I go to the track less than five times a year, and so cutting this activity out of my life would save, at most, 50 gal of fuel annually. Now let's put this into perspective: in 2014, Americans alone used 136.78 *billion gallons* of fuel (U.S. Energy Information Administration, 2015). That means that my fuel consumption for this activity constitutes a mere 0.00000000037 % of *just America's annual fossil fuel use*. Of course, the rest of the world burns fossil fuel as well, and fossil fuel is not the only source of GHG. In other words, my yearly contribution to global emissions through this activity is *infinitesimal*; it approaches zero. This seems like a plausible case of my activity making a truly insignificant difference.

Procreation, however, isn't quite like taking my motorcycle to the track. It isn't simply that creating another person has immediate, high-emissions consequences (although any parent will tell you that it does this as well!—after all, you must buy diapers, wash extra clothes, sometimes move to a larger home, buy a larger car, purchase more food, etc.); in addition, procreating has a *carbon legacy*, in that there is now a new person, who will become a consumer and emitter in her own right, and potentially make even more people.

This insight led Paul Murtaugh and Michael Schlax (2009) to attempt to calculate the emissions impact of having a child. Their motivating question was: if parents are responsible for even some of their offsprings' emissions, what might a total accounting of the environmental impact of procreation look like? And their findings are staggering. For the sake of comparison, Murtaugh and Schlax chose six common, praiseworthy, 'green' activities and calculated their lifetime emissions savings. These activities were: increase one's car's fuel economy from 20 to 30 mpg; reduce miles driven per week from 231 to 155; replace traditional windows with energy-efficient models; replace ten 75-w incandescent bulbs with 25-w, energy-efficient bulbs; replace one's old refrigerator with energy-efficient model; and recycle newspaper, magazines, glass, plastic, aluminum and steel cans. They then compared the emission savings of these activities with the emission savings of refraining from having a child, under various emission projections for the coming generations. Under a constant-emissions scenario—in which we continue along 'business as usual' and individually maintain the current average annual emissions, the lifetime emissions savings of choosing not to have a child is *more than 20 times* that of the above six activities *combined* (2009, p. 18). Further, even on a much more optimistic scenario, in which we immediately adopt the IPCC's guidelines for decarbonization, and so future generations radically reduce emissions and eventually become net-zero emitters, the choice not to have a child *still* resulted in a higher emissions savings than the cumulative lifetime totals of all six 'green' activities (2009, p. 18).

Another comparison to help us see the fairly radical effect that procreation has on one's emissions is by comparing it to one's lifetime, non-procreative emissions. According to Murtaugh and Schlax's calculations, the fact of carbon legacy—that is, the fact that one's children will go on to live and emit, and perhaps procreate themselves—results in the rather strange implication that the activity of having a child raises one's lifetime carbon emissions *by several times*. In particular, on the same constant-emissions scenario, each child that an individual has adds about 9441 metric tons of carbon dioxide to her carbon footprint, which is *5.7 times the lifetime average emissions of an American's non-procreative activities* (2009, p. 14).

Most people are shocked by these numbers; but on reflection, we shouldn't be. For as long into the future as humans are GHG emitters, our offspring will be continuing to make the problem worse. So while my engaging in other high-emission activities (like taking a trans-Atlantic flight, for instance) have negative effects for the environment, these are one-time costs. When I procreate, I stand on top of an iceberg of future emissions as my family tree branches into the future. And these emissions don't stop unless and until we fully decarbonize our economy such that each future individual is a net-zero emitter.

This sounds bad for procreating. In the context of other actions that we can take to curb our emissions, having a child is in a class by itself. And the way I have described it here, the carbon effect of having a child seems *massive*. However, even on the constant-emission scenario explored by Murtaugh and Schlax—which we certainly hope is a worst-case-scenario—having a child 'only' results in 9441 metric tons of carbon dioxide. And while that number seemed very large in the context of other actions one could take to mitigate her carbon footprint, or even in the context of one's average, non-procreative carbon footprint, it is not large in the context of global emissions. Remember that number from earlier? Globally, we emit more than 30 billion tons of carbon dioxide each year. Or, to zoom out even further: the all-time anthropogenic carbon budget—the amount that we can emit before raising the global average temperature by 2 °C—is about a trillion tons. In the context of these numbers, even a high-emission activity like procreating seems to have an infinitesimal effect. Having a child, or two, or even ten, doesn't seem to make a significant difference to climate change, given the scale of the problem.

2.4   Absolute and Relative Significance

There was *something* important about the intuition that procreating has a significant environmental effect, though. While focusing on global emissions or the all-time anthropogenic carbon budget makes the emission effects of procreating seem insignificant, focusing on the effect one can have through other individual activities makes the environmental effects of having a child seem quite significant. So what is going on here?

The problem that we are discovering is that the language of significance is vague. On the one hand, we can read 'significant' as an *absolute* or *all things considered* modifier; when we have the intuition that taking a pleasure drive, even in a Hummer, is not significant, we are likely employing this absolute sense. In the grand scheme of things—considering the total anthropogenic GHG emissions—it is obvious that the emissions from any single action that I could take are insignificant.

However, we also regularly use 'significant' in a *relative* sense, as we might when we compare the emissions of the Hummer to that of a hybrid sedan. If driving a hybrid would cut my emissions by, say, 70 % over that of driving a Hummer, we might well think that the emissions of the Hummer are significant, *relative to the emissions of the hybrid sedan*. When we see the carbon legacy of procreation, as in the Murtaugh and Schlax study, it is in this relative sense that the emissions effect of procreating may strike us as significant. Procreating swamps all of our non-procreative activities in terms of its emissions effects; there is no other single act that you can refrain from that will have anything like the environmental impact of refraining from having a child. That is: relative to any other possible action one can take (or indeed, to all possible actions one might take), procreating is environmentally significant.

This relative sense of significance was not, alas, the sense employed in **Significant Difference**. It is in reading significance as absolute that **Significant Difference** seems compelling. The question implicitly asked by such a principle is: why would I be obligated not to act in a way that doesn't have an all-things-considered significant effect on the problem that threatens the morality of the act in the first place? So the language of significance is vague, in that it can be read in both an absolute and a relative sense; and **Significant Difference** employs significance in the absolute sense, while the argument that made procreating seem significant employs significance in a relative sense. It thus looks like the argument concerning the *relative significance* of procreation's environmental effects does not undermine **Significant Difference**.[2] Although we will see the idea of relative significance re-enter the discussion later, for now it appears that a successful argument from the threat of climate change to procreative obligations must refute **Significant Difference**. I turn to the initial stages of that task now.

---

[2] Even given the disambiguation that I attempt here, one might resist this conclusion. One might, that is, argue that the act of procreating does approach making a significant difference, and that is because we should think not only about the scale of the problem (that it takes a population of 7.3 billion emitters to cause the harms of climate change), but also the scale of the harms (millions— perhaps even billions—of people will be harmed by climate change). On this view, we can calculate the 'statistical harm' that one does by emitting, and when doing so, we will note that raising one's lifetime emissions by several times makes a significant difference to this statistical harm.

This sort of argument has been made, for instance, by philosopher John Nolt, who argues that as a result of her lifetime emissions, the average American is responsible for the suffering or death of one to two future people (Nolt, 2011). If one were convinced by this argument, then raising one's lifetime emissions by, say, six times, would amount to being responsible for the suffering or death of an additional 6–12 people, and surely this would be significant.

However, Holt's calculations are (as he admits) crude, and there are reasons to be suspect of the entire notion of statistical harm. In addition, as I will note later in Sect. 3.1, it is not only the scale of climate change that makes responsibility for harm difficult to attribute—it is the *complexity* of the problem. Thus, for the sake of remaining as modest in my conclusions as possible, I will not adopt this framework. Needless to say, if one is tempted by Holt's reasoning, then the challenge from **Significant Difference** is met immediately, and there are reasons to think we might have procreative obligations in addition to whatever my arguments establish here.

2.5 Non-Consequentialist Intuitions About Significance

Significant Difference is a *consequentialist* principle, as it assumes that what determines the moral status of some action is the consequences of that action. Now, it's a fairly refined consequentialist principle, since the consequences count in a subtle way: it's not that one is permitted to act when acting makes *no* difference; rather, one is permitted to act when acting makes no *significant* difference. But what is key is that, according to the principle, *only* a significant difference in contribution to a moral problem could justify requiring that I not take my motorcycle to the track for fun. And *this*, I want to point out, is a contentious claim.

In moral theory, many philosophers have pointed out that purely consequentialist views are often unsatisfying.[3] One way in which they are unsatisfying is their inability to explain cases in which our intuitions seem quite settled. For instance: must I recycle my trash rather than throwing it away, when the two bins are right next to one another? It seems, to me at least, like I must, despite the fact that failing to recycle my trash doesn't make a significant difference with regards to any of our environmental problems. What seems relevant is that the environmental cause is a just one, and by recycling I am doing my part. So despite the fact that this individual contribution makes no significant difference, it seems like I am obligated to recycle. This same reasoning would require that I turn out lights when I leave a room, utilize energy-efficient appliances, and institute various other 'green' practices.

Now, certainly, I owe an argument, or at least an explanation for what might ground an obligation in these cases. Were my actions to make a significant difference, we would likely say that I have an obligation not to make significantly worse some very serious moral problem. But that can't be our justification in this case. So what moral principle(s) might justify these non-consequentialist intuitions? John Broome has helpfully built on the common distinction between duties of *goodness* (which concern making the world better) and duties of *justice* (which concern what we *owe* to particular others, regardless of whether it makes the world better) in the context of climate change. Broome argues that our *individual moral burden* comes from duties of justice, precisely because no one of us can, by herself, make a significant difference to the problem of climate change, whereas our time and resources can make a significant difference to other moral problems (such as poverty alleviation and disease treatment (Broome, 2012, pp. 64–68)). *Institutions*, then, are the primary bearers of duties of goodness, since they do have the ability to make

---

[3] For a small sample of such concerns, consider the following: consequentialism seems unable to account for *justice*, as acting in paradigmatically unjust ways (such as framing an innocent man, or enslaving a minority population) may turn out to best promote the overall good; some forms of consequentialism also seem to 'fail to take seriously the separateness of persons', in that the good for one may be sacrificed for the good of others, as if each individual were only a part of one larger individual (Rawls, 1971); many forms of consequentialism seem overly 'demanding', or seem to threaten the *integrity* of a moral agent (Williams, 1973). There are many others, and of course, consequentialists believe that they have responses to all such worries. But for present purposes, it suffices to note that a principle's being consequentialist makes it susceptible to several kinds of serious, theoretical concerns.

significant differences.[4] On his view, then, Sinnot-Armstrong is correct in a sense, in that none of us has an individual moral obligation of *goodness* to combat climate change; his addition, then, is simply that there is another kind of obligation (that of justice), and we do, in fact, have that sort of duty.

Broome's particular solution is interesting, as he thinks that each of us has a strict duty not to emit carbon (period!) as a matter of justice, but that this is easy to accomplish for most of us through the purchase of carbon offsets (in this way, each of us can be *net-zero* carbon emitters). I will not, in this book, accept his particular solution, for a variety of reasons. Practically, this is because, as we saw in the discussion of *relative* significance, procreation is very carbon-expensive. As a result, it is less obvious whether most people could afford to offset their procreative activities, or even how this could be conceptualized (since procreation happens at a time, but the costs of it are distributed over generations).

More importantly, however, is that offsetting is not a 'magic bullet'. While there is a sense in which offsetting one's carbon emissions results in her being a 'net zero' emitter, there is an important sense in which this is misleading. Offsetting works by having the emitter pay the cost of removing the emitted amount of carbon from the atmosphere. Most commonly this is done through paying for the creation of a new 'carbon sink', such as planting a tree, or the protection of an existing carbon sink, such as an existing tree. The idea, then, is that through offsetting, one emits whatever carbon she does, but then ensures that the same amount of carbon is either taken out of the atmosphere or prevented from being emitted elsewhere. Thus the language of being a 'net-zero' emitter.

While offsetting is clearly a good thing to do, it is easy to see why it is not a perfect solution. The sort of carbon sinks that can be promoted through use of one's money (trees, marshes, etc) are *short-term* carbon sinks. If I plant a tree, then it will absorb CO_2 from the atmosphere and photosynthesize it into O_2; it will do this especially for the years that it grows into a mature tree. But within a few decades or centuries, the tree will die, fall, rot, and eventually decompose, releasing the trapped CO_2 back into the atmosphere. Although this certainly doesn't mean that we shouldn't be concerned to protect and promote the earth's short-term carbon sinks, the important point here is that the scale of these carbon sinks is drastically different from the long-term carbon sinks of coal and fossil fuels. It takes *millions of years* for carbon to be trapped in deep carbon sinks, and so burning fossil fuels liberates

---

[4]C.f. (Broome, 2012, pp. 50–54). Broome's particular view is that our duties of justice in the case of climate change require that we not harm any individual, and that any emissions at all constitute harming others (especially the worst off). His view, then, is that duties of justice often co-travel with duties of goodness: we are obligated not to harm particular others, and this has a goodness justification as well as a justice justification. However, Broome motivates the distinction by noting that justice obligations are those that step in to normatively require that we act in certain ways, even when doing so fails to prevent the better outcome, or even when doing so actively brings about the worse outcome. So one can (and I will) abandon Broome's focus on the duty not to harm, since one of the great insights of the goodness/justice distinction is that obligation can come apart from harms and benefits.

CO_2 from a sink to which it cannot be returned on a human timescale. The carbon sinks that we can manipulate are a short-term fix for a very long-term problem.

So I have to disagree with Broome that offsetting one's emissions is all that duty requires. It is clearly a good thing to do, but it does not make us truly 'net-zero' emitters. However, the framing of duties of goodness vs. duties of justice is a good one, and I adopt a version of it here. As a result, I will, in Chap. 3, continue to justify my rejection of **Significant Difference**, exploring various plausible moral principles that seem to imply an obligation to act in ways that may not make a significant difference, and so justify candidate procreative obligations. In addition, since I am interested in more than *strict obligation*, I will later consider other moral concepts as well, which seem to imply some weaker degree of moral burden on individuals.

2.6 Conclusion

In this chapter, we have looked at a common problem in attempting to allocate individual moral responsibility as a result of massively collective crises. In particular, we have discussed the objection from causal impotence, or what I eventually formulated as the principle of **Significant Difference**. This sort of objection is common in the climate change literature, because the challenge of reducing our GHG emissions is so massive that most individual actions make virtually no difference to the problem, and so intuitively, it seems strange to think that one might be prohibited from those actions.

What seems plausible is that, if **Significant Difference** is true, then we have no obligation to refrain from having any number of children. No single activity that a person can take makes an *absolute* significant difference to the extent or severity of the harms of climate change, even though procreation is certainly *relatively* significant in its carbon costs. Thus, if we are to have any procreation-limiting obligations in light of climate change, it must be because **Significant Difference** is false. In the previous section, I have suggested that many of us likely have a set of intuitions indicating that this is the case. For very many of us will think that we can be obligated to recycle, for instance, even though our recycling makes no significant difference to any environmental problem. If we are right about this, then **Significant Difference** must be false, and there must be, instead, some principle or moral justification requiring action even in the face of causal impotence. In the following chapter, I will explore several candidate justifications for such action.

References

Broome, J. (2012). *Climate matters: Ethics in a warming world*. New York: W.W. Norton.
Jamieson, D. (2014). *Reason in a dark time: Why the struggle against climate change failed—and what it means for our future*. Oxford: Oxford University Press.

Murtaugh, P. A., & Schlax, M. G. (2009). Reproduction and the carbon legacies of individuals. *Global Environmental Change, 19*, 14–20.

Nolt, J. (2011). How harmful are the average American's greenhouse gas emissions? *Ethics, Policy and Environment, 14*(1), 3–10.

Rawls, J. (1971). *A theory of justice*. Cambridge, MA: Belknap.

Sinnot-Armstrong, W. (2010). It's not my fault: Global warming and individual moral obligations. In S. M. Gardiner, S. Caney, D. Jamieson, & H. Shue (Eds.), *Climate ethics: Essential readings* (pp. 332–346). Oxford: Oxford University Press.

U.S. Energy Information Administration. (2015, March 12). *How much gasoline does the United States consume?* Retrieved January 29, 2016, from eia.gov: http://www.eia.gov/tools/faqs/faq.cfm?id=23&t=10.

Williams, B. (1973). Integrity. In J. Smart & B. Williams (Eds.), *Utilitarianism: For and against* (pp. 108–117). Cambridge: Cambridge University Press.

Chapter 3
Individual Obligation

In the previous chapter, I mentioned that John Broome, in his discussion of obligations regarding climate change, borrows a helpful Kantian distinction between duties of justice and duties of goodness. On his view, recall, institutions—which are able to make massive changes to emissions—are the primary bearers of duties of goodness, as individual emitters simply do not do significant good by directing their resources towards mitigating climate change (Broome, 2012). If the arguments of the previous chapter are on track, then Broome's argument appears to support the causal impotence objection: given the scale and complexity of climate change, virtually nothing an individual does could matter to the climate-related outcome. While each emitting activity makes a technical difference—resulting in slightly more atmospheric carbon than there was before—it does not make anything like a significant difference to the overall problem of climate change.

What Broome pointed out, though, is that even though we do not, as individuals, have duties of goodness regarding our emitting behavior, there are other candidate duties that individuals could bare. He calls these 'duties of justice', but this title can be slightly confusing, for a couple of reasons. First, 'justice' considerations are often taken to be more specific than merely the counterpart to 'goodness' considerations (in particular, justice is often taken to be related to 'fairness', 'desert', or 'equality'); so, for instance, *one* candidate principle below is a principle of justice, but there are others as well. And second: in exploring ways that each of us may have a moral burden or responsibility regarding our procreative behaviors, we might think that there are considerations other than 'duty'. Indeed, later in Chap. 5, I will borrow from Dale Jamieson the language of 'Green Virtues' to articulate the idea that perhaps we ought not to see ourselves as obligated to act in a certain way, but rather that we ought to develop certain character traits or virtues that predictably lead to our adopting environmentally-friendly practices. But having a virtue does not necessarily entail having any particular obligations. For these reasons, then, I will not discuss 'duties of justice' as the counterpart to duties of goodness; rather, I will first explore candidate non-consequentialist principles that seem plausibly to

© The Author(s) 2016
T.N. Rieder, *Toward a Small Family Ethic*, SpringerBriefs in Public Health,
DOI 10.1007/978-3-319-33871-2_3

generate individual duties. Then, in Chap. 5, I will broaden the discussion further to consider other sorts of 'private moral burden'.

The goal of this chapter, then, is to propose three principles as plausible candidates for generating private procreative obligations in the context of climate change and the population crisis. That is: I will present three moral principles that would seem, given the facts about our climate and the global population, to have *some* implication for our procreative practices. The precise content of our procreative duties, however, will be a live question. Thus, in Sects 3.1, 3.2 and 3.3, I will simply present the candidate principles and argue for their plausibility. Only then, in Sect. 3.4, will I ask what they might entail specifically for our procreative behaviors. If any of the proposed principles is valid, then there seems to be good reason to believe that we ought to *restrict* our procreative behaviors; but does that mean that each of us is obligated to have no children? Or simply not too many (whatever that might mean)? Although I do not believe that there is an obviously correct answer to this question, I will suggest a sort of 'limit' to an acceptable answer.

Regardless of the particular content of the duty, however, the upshot of this chapter is that we may, in fact, have *some* obligations to limit our procreative behavior. That is: even if we have no duty of goodness to limit the number of children we have, there is still disconcertingly powerful reason for thinking that our procreative activity is subject to the demands of duty.

3.1   Duty Not to Contribute to Harms

The first candidate moral principle is a duty closely related to the duty not to harm. If it were the case that emitting carbon dioxide directly and obviously harmed, then there would be no problem making the case that we have a duty not to emit carbon dioxide (or to restrict our emissions in some way).[1] However, the first problem with utilizing such a principle was investigated in the previous chapter: the harms of climate change are the result of a massive collection of unrelated acts by uncoordinated individuals, and so it actually seems wrong to say that an individual act harms; this is why we focused above on the notion of 'making a difference' to the extent or severity of climate change. So in the context of a massively collective action that harms, we might think that our duty is not to make a significant difference. We discussed that candidate last chapter, and I am proceeding under the assumption that an individual act of procreation—like our other climate-related acts—makes no significant difference to the harms of climate change.

There is yet another problem with appealing to harm, however, and that is the *complexity* of the climate system, and the way in which our small, individual contributions of GHG get diffused throughout a massive system. Much of my individual emissions, for instance, may end up in a natural carbon sink, just through accident,

---

[1] Recall that this is how Broome actually gets to his conclusion that each of us is required to be a 'net-zero' emitter.

in which case my particular emissions didn't even causally contribute to the harms of climate change (since my emissions aren't warming the atmosphere). This radical complexity and uncertainty leads some ethicists, like Dale Jamieson, to claim that not only do we not harm anyone with our emissions, but we don't even partially *cause* the problem with our emissions, or reliably and predictably *raise the probability* of climate harms by our emissions (Jamieson, 2014, pp. 144–169).

This issue of causation is exceedingly difficult, and one might be skeptical that one's emissions, small though they are, play *no* causal role. After all, even if I get lucky, and my emissions get taken out of the atmosphere by a natural carbon sink such as a forest or the world's oceans, my emissions have just used up a small fraction of the earth's available carbon sinks, displacing other emissions into the atmosphere. In addition, not all ways of removing carbon from the atmosphere are equal; the forest that absorbs my CO_2 is a relatively short-term carbon sink, and the death of the trees in the future will release the gas back into the atmosphere; and the ocean is becoming more acidic as it absorbs more carbon dioxide.[2] So if I burn fossil fuels, I have liberated CO_2 from a long-term carbon sink; as a result, even if it gets removed from the atmosphere, it may displace other people's emissions from a carbon sink, or end up in the world's oceans, which acidify as they absorb carbon dioxide. In both cases, we might think that the very act of liberating plays *some, minute* causal role in the overall climate change problem.

But do these various fates of my emitted carbon dioxide constitute partially *causing the harms* of climate change? Again, the issue is clearly difficult. We would need a sophisticated account of causation, and any answer given would be subject to reasonable challenge. However, I don't think that we must focus on our causal role in harming in order to understand how we might have a duty not to *play a role* in the problem that causes harms. What could playing a role mean, if not partially causing the problem? Let's take a look.

There are a few different ways that we might think someone is playing a role in a serious moral problem, even if it was unclear whether her acts partially cause the problem.[3] One way might be acting in a way that would otherwise be innocuous, but which one knows produces something that is part of a massively problematic system. Consider the example of a low-level researcher who does basic science for a terribly corrupt corporation or political regime that uses all of its resources to harm innocent people. Given the kind of science she does, it is not the case that the scientist will produce a bomb or other mechanism of destruction for her tyrannical bosses; but she *is* producing something—knowledge—which will become part of a terrible system and which, through some convoluted and unpredictable causal

---

[2] Increased atmospheric CO_2 has led to oceans becoming about 30 % more acidic than they were prior to the Industrial Revolution. According to business as usual predictions, we may see a further 150 % rise in acidity by the year 2100, which would bring oceans to a pH level not seen in more than 20 million years (National Oceanic and Atmospheric Administration, n.d.)

[3] This line of thinking was originally inspired by Fruh and Hedahl (2013); the following explication of various ways that one can 'play a role' in systematic harms somewhat parallels that described in (Hedahl, Fruh, & Whitlow, 2016).

system, may someday help (in some very small way) the evil regime to do something awful. We can call this *contributing* to a system that harms.

In a different case, we can imagine German citizens during the Nazi occupation who are told to salute and chant, 'Heil Hitler' at various times. It is exceedingly implausible that honoring the Nazi regime in this symbolic way actually harms the Nazi's victims; however, doing it makes one an active participant in the abhorrent regime. It may well be that, given the particular costs of defecting in this case, it would be all-things-considered permissible to do as the Nazi's demand. But the moral reaction we have suggests that there is, in fact, a problem, and it is with the role that we play regarding the morally awful system. Call this a case of *participating* in a system that harms.

Finally, there is the even more standard case of standing idly by while massive harms are being perpetrated. It is likely that in the United States, prior to the Civil War, there were at least some individuals who understood the moral horror of slavery, but who said and did nothing about it. These individuals would have benefitted from the practice of slavery—buying cotton products and food at lower prices thanks to slave labor—but would not have harmed any of the slaves themselves. The moral disturbance in this case doesn't come from the causal role in harming—it comes from the failure to fight an injustice, especially when the injustice provides one with benefits. In this case, we might think that an individual benefitting indirectly from the practice of slavery is *complicit* in its massive, systematic harms.

Contributing to, participating in, and being complicit in massive, systematic harms all seem morally bad, but to varying degrees. Perhaps one is not obligated to avoid complicity in all harms, but that complicity generally reveals cowardice or other vices. And perhaps participating in a system of harm is more objectionable, but still understandable and even excusable if the costs of failure to participate are very high at all (as in the Nazi case). The case of contributing to a massive, systematic harm seems the worst, as the role that one plays is more significant; it may still be a stretch to say that such a person *caused* any particular harms, even partially; but she did actively *contribute* to the system that did the harming. Further, it's worth noting that it's not always easy to distinguish between these different ways of playing a role in systematic harms, and that there is likely significant overlap; indeed, *contributing* to a system of massive, systematic harms will likely typically include *participating* in that system and being *complicit* in its harms.

It seems plausible to me that each of the ways of 'playing a role' in massive, systematic harms discussed above is plausibly *prima facie* morally wrong.[4] Our

---

[4]Literally 'on its face', the language of *prima facie* was adopted by philosophers to denote the provisional character of duties that have not yet been weighed against the competing goods of the actual world. A *prima facie* duty, then, is one that I am required to follow, if it is not outweighed by some other consideration. *Prima facie* duties are contrasted with *all-things-considered* duties, which emerge at the end of the weighting and balancing process among the various, relevant goods and reasons, and which tells us what we must, in the end, do. While some *prima facie* duties seem to always imply an all-things-considered duty ("do not murder," for instance), others are so all-encompassing that they regularly admit of trade-offs ("promote the good," perhaps). What we seem to be learning at this point is that the duty not to contribute to massive harms seems to be

moral reactions to each of the cases are, I think, evidence that the action in question violates a duty—a duty not to play a role in massive, systematic harms. However, I did admit that complicity seems less bad than participation, which seems less bad than contribution to systematic harms. So I will formulate my candidate principle in the weakest way possible, and suppose only that there is a **Duty Not to Contribute to Massive, Systematic Harms**. This is not a duty not to *cause* harm—even partially—but is rather a duty not to inject oneself as an active contributor into the large, causally complex machine that is doing the harm.

This duty would make sense of how we judge many acts that either don't make a significant difference to a moral problem, or don't partially cause a serious moral problem at all. The example from the previous chapter was recycling: it seems I am obligated to throw my waste in the recycle bin rather than the trash can, even though my throwing a single piece of refuse into the trash would not make a significant difference, and would not clearly cause any harm. The justification is that waste management is a massive moral problem, and by throwing away my trash, I am contributing to it. Those who oppose factory farming might make a similar argument for the duty not to buy certain meats. Although some philosophers have argued that the very small causal role that one plays in the continued harm of animals justifies the duty not to purchase meat (see, for instance (Norcross, 2004, pp. 232–233)), the causal complexity of the system of factory farming might make the principle under investigation seem to be a more plausible justification. Factory farming is a system that generates massive harms for sentient creatures, and so we have a duty not to contribute to that system, and our small marketplace exchange is a form of contribution.

The duty not to contribute to massive, systematic harms makes sense of many of our environmental obligations, even if most individual activity does not make a 'significant difference' to the extent or severity of those harms. Now, it is merely by bringing the act of procreation into the discussion of emissions activities that we see its implications for our procreative behaviors. If a duty not to contribute to harm can ground my duty to recycle, then surely it could ground an obligation to limit our much more carbon-expensive activity of making babies. Of course, to what extent our procreative activities must be limited is, as I said in the introduction, an interesting question in itself, and so I will come back to it after having discussed the other candidate principles.

3.2 Duties of Justice

Another possible consideration in favor of procreative restrictions is that of justice. Now, the language of justice is admittedly broad, and we have already seen one way in which different concepts may be intended: Broome discussed duties of justice as

---

more like the latter than the former, and so discussion of its relative justificatory burden is important.

counterparts to duties of goodness, in which case very many possible duties may be considered duties of justice. However, we often invoke justice to mean something more particular, concerning fairness and various kinds of equality.

When I refer to justice, I intend this narrower sense of the concept. And while I cannot here provide a particular theory of justice, I will employ the language of justice to cover considerations of fairness and some degree of equality among all persons. My hope is that this will allow me to suggest what seem to be plausible moral principles, but while maintaining a level of abstraction that prevents theoretical in-fighting.

In some ways, the demands of justice are the easiest to describe. The grounds for thinking that justice applies to the procreative context are simple, and the language of fairness and equality are intuitive. I will suggest two, related grounds for thinking that justice might issue demands on our procreative behaviors.

Firstly, overpopulation is a problem that disproportionately harms the poorest and most vulnerable of the world's population, even while their procreative activities contribute least to the problem. That the world's poorest are most harmed is easy to see: as the world's resources become depleted, and as climate change worsens, it is not the wealthy elites of the first-world who will suffer first. It is the poorest residents of Bangladesh who will simultaneously deal with food shortages, lack of access to fresh water, and increased incidence of devastating storms and flooding (Intergovernmental Panel on Climate Change, 2014, Chap. 24: Asia). It is the island inhabitants of the Maldives and Kiribati who will lose their homes to rising sea water. And it is those without access to sanitation and health care who will be most affected by changing disease vectors.

Regarding their contributions to the problems of overpopulation, one might be surprised to see the claim that the poor and vulnerable have contributed least; after all, the fertility rate of wealthy nations is typically between 1.5 and 2.0, while the fertility rates of poor regions of Africa, Asia and the Middle East regularly approach 6.0–7.0 (United Nations, Department of Economic and Social Affairs, Population Division, 2015). Doesn't this suggest that, although the poor may suffer early and badly, that they also have largely contributed to the problem?

In fact, the answer is 'no'. The problems of resource use and climate change are problems that depend not only on the number of people contributing to the problem, but also their levels of contribution.[5] The high fertility rates of West Africa, then, do

---

[5] We should be careful to recall from the introduction, that there are many resource-related reasons to be concerned with overpopulation, and I have chosen to focus on only one of them—climate change. So it could be argued that the poor residents of the world who have five, six or seven children are still contributing more to *overpopulation* than most Americans are, even if that overpopulation isn't as relevant to the particular problem with climate change. However, all of the problems with overpopulation have a similar structure as the one I am dealing with here: it is not the sheer number of people that is problematic, it is the number of people given the limited availability of some resources (clean water, food, energy, etc.). And so it is actually quite difficult to argue that the high-fertility-rate poor population is contributing to the problems of overpopulation in *any* way, since they consume so few of the available resources. Thus, while I deal explicitly in the main text only with the case of climate change, the reader may choose to pursue another issue of resource

not necessarily entail large contributions to the problems of overpopulation, as their citizens use a fraction of the resources that, say, American citizens use. Take Niger and the US as an illustrative example: although Niger has the highest fertility rate in the world, at 7.6, and the US has a relatively low rate of 2.1 (United Nations, Department of Economic and Social Affairs, Population Division, 2015), the variation in the two countries' *per capita* CO_2 emissions is even more staggering. The average American emits around 17 metric tons of CO_2 per year, while the average Nigerien (not to be confused with Nigerian) emits an astounding 0.1 metric tons of CO_2 per year (The World Bank, 2011–2015). The average US citizen thus emits nearly *200 times* the amount of CO_2 of the average Nigerien, and so the average procreative behavior of an American (having two children, who together will emit around 34 tons of CO_2/year) is vastly more damaging to the problem of climate change than the average procreative behavior of a Nigerien (having seven children, who together will emit a mere 0.7 tons of CO_2/year).

The first point, then, is that the world's wealthy do much more to contribute to the problems of overpopulation than the world's poor and vulnerable, and yet the poor and vulnerable will be harmed the worst. This situation should strike us as *unfair*, and is a central violation of what ethicists often call *social justice*. If anything seems clearly true in the realm of justice, it's likely that this structure is morally bad, and so we have one reason for worrying that procreative activities are subject to the demands of justice. Before asking exactly how, let's consider the second, related, way of formulating the justice concern.

Recall from Chap. 1 the models of population sustainability. The primary upshot of that discussion was that the Earth cannot sustain a population of 7.3 billion wealthy citizens—and in fact, cannot likely even sustain our population under the current distribution of wealth. We are in an ecological 'overshoot', in which we are using 60 % more resources each year than would be sustainable. So what does this fact entail for my procreating?

Well, given that my child will be an American, I can predict that she will use an incredible amount of resources—an amount that could not be used by each inhabitant of earth. What this means is that my having an American child, against the backdrop of overpopulation, *depends on the abject poverty of others*. When I create a wealthy, resource-expensive person, I am doing something that *requires* either a smaller population, or a radically unequal distribution of resources. Another way to put the point: my having an American child is *incompatible* with very many others doing the same. And this, too, seems unfair.

Justice, then, raises two, related worries for procreation by the global wealthy: by making a new person, I both (1) contribute relatively largely (relative to those being harmed, not relative to the scale of the problem) to a problem that will harm the world's poor first and worst; and (2) create the kind of person whose existence depends on a radically unequal distribution of resources. On its face, then,

---

shortage on her own, in order to see whether the populations of, say, the poorest countries in West Africa, might really be contributing to the *problems* of overpopulation, given these peoples' lack of access to resources.

considerations of justice look to condemn unfair levels of procreation by the world's wealthy. Again: what 'level' of procreation is unfair is a difficult question, on which we will hold off until after discussion of the final candidate principle.

3.3   Obligations to Our Possible Children

Thus far, we have investigated possible procreative obligations as a result of how procreation affects unrelated *others*; by procreating, we contribute to a system that will cause massive harm to millions of others—in particular, to those who are already badly off. The kind of reasons that such concerns provide one with are 'agent-neutral', in that they provide the same reasons to everyone. The fact that climate change will drown the islands of Kiribati and the Maldives provides everyone with the same reason not to make (especially carbon-expensive) new people.

However, the facts of overpopulation and climate change may justify yet another sort of duty, as a result of the effects on the children that one does have. In short: the dire moral threats of today lend real credence to the classical cynic's worry that being brought into existence may not be in one's interest. This sort of moral worry, however, is *not* agent-neutral: it concerns *my* child, and invokes the obligations that I have as a result of becoming a parent. According to this worry, then, I may have special moral reason to protect my own children from living lives that involve certain kinds of harms. In the current section, we will investigate the case for the existence of 'agent-relative' reasons not to procreate.

In a controversial essay in *The New York Times*, philosopher Peter Singer raised the question of whether the current generation should be the last one. His own answer is 'no', that it should not, because he thinks a world with human life on it is better than one without (Singer, Last Generation?, 2010). However, he made vivid the pessimist's worries that climate change, overpopulation, and the myriad of other threats to human life make bringing a child into this world distinctly risky for that child. What kind of life is my child likely to live? Can I provide my child—the being that I will predictably love the most, and most want to protect from harm—anything like the kind of life that I would want for it? In a follow-up article, Singer quotes a thoughtful commenter, who expressed the worry well, saying, "I love my children so much that I didn't have them" (Singer, Response, 2010).

On the one hand, the idea here is clear: the problems investigated in this book—climate change and overpopulation—in addition to many problems not here discussed (such as the threat of superbugs, nuclear war, terrorism, etc.) make the prospects of a good life for future people seem increasingly dim. All of these threats seem to make it likely that any child I have will suffer serious harms. But, of course, as a parent I will want to protect my child from harm. And so, perhaps the correct course of action is to protect my children in the only way that is guaranteed to work—namely, by not creating them.

This worry is certainly not new. Cynics and misanthropes have long wondered whether we are cruel to bring new people into this terrible world. But the catastrophic and global threats of the day—in particular, of climate change—lend what may otherwise be seen as a fairly unserious, if wry, commentary on our world rather more credibility. If, as seems virtually inevitable, the next generation will see global average temperatures rise at least 2 °C (and perhaps as much as 4 °C, if we do not act swiftly), the world will become a distinctly worse place, and the population will suffer. Does that, in fact, make it cruel to have a child? Does one have an agent-relative reason, based in the obligation to protect one's child, not to procreate?

In fact, the question is philosophically quite complex. The reason for this complexity is that whether an obligation to protect one's child applies to the procreative context requires us to determine whether or not coming into existence can be a *harm*. After all, if it can't, then there would be nothing that a parent need protect her child from. So: can coming into existence be a harm?

A quick argument claims that the answer is 'no'. This is because we tend to think that the concept of 'harm' (and the mirror concept of 'benefit') is *comparative*—that is to say, that it requires a comparison between two states. So I am harmed only if I am taken from a state of relative goodness to a worse state; and vice versa with benefitting. This seems to be how we use the concepts in ordinary contexts: I am harmed if you hit me with a bat, because you moved me from a state in which I did not have a head injury into a state in which I do have such an injury.

The problem with claiming that procreating can harm, then, is obvious: by procreating I do not move my child from one state into a worse state; indeed, I do not move my child between two states *at all*. This is because *non-existence isn't a state that one can be in*. It is the lack of having a state at all! This very sort of consideration has been used to justify the creation of a child with disabilities, even when the parents could have avoided it, since being created with a disability simply is not a harm. It may be worse than being created without a disability, but that doesn't imply that creating such a person harms them. This counter-intuitive implication is often referred to as 'The Paradox of Harm'.

So at first glance, it looks as though these very strange, abstract, philosophical considerations entail that one cannot harm her offspring by procreating, and so no duty to protect one's child could generate a reason not to procreate. But that would be to move too fast. For in response to The Paradox of Harm, one might wonder whether it's really plausible that *no* procreative acts can really be thought to harm the created individual. What about a child who is born into a life full of massive, debilitating suffering, who lives in this state for a few months and then dies? Such a child's short life is nothing but misery; is it really plausible to hold that creating such a child does not harm her?

In response, some philosophers claim that such cases do show that procreation can harm the child created, but *only if the child's life is not worth living*.[6] In the case

---

[6]This is the structure of Derek Parfit's oft-cited discussion in (Parfit, 1984). There, Parfit was concerned primarily with what he called the 'non-identity problem'; however, in setting up the problem, he noted that it doesn't seem we can harm a child by creating it, unless its life would be not worth living. The non-identity problem, then, is the tension between our belief that one perhaps ought to wait to have a child, if doing so would result in a healthier child, and the fact that such

of a short, miserable life, what we judge is that existence is worse than non-existence, and so our comparative notion of harm actually applies: by creating such a person we make her worse off, and so harm her. In such radical cases, the parents may, in fact, be given a reason not to create such a child by the general duty to protect one's offspring.

If the above is the correct way to think about harm and benefit in the procreative context, then very many of us likely are not given reason not to procreate by the dangers of climate change. After all, I—and likely anyone reading this—can expect that any children we have will have lives worth living, even if they would be worse than we might ideally want them to be. But if their lives will be worth living, then existence is not worse than non-existence, and so creating them does not harm them.

I want to suggest, however, that the question of whether we harm or benefit our children in creating them is not particularly helpful. I said earlier that the difficulty of thinking about harm and benefit in the procreative case is that, prior to procreation, one's child does not have a situation to be changed; non-existence is not a state. But that makes the above diagnosis of 'having a life not worth living' as being what harms a child seem very strange. After all, if non-existence is not a state one can be in, then just because one's life is so bad as to be not worth living does not mean that creating such a person moves them from a better state to a worse state; we just said that non-existence isn't a state. So if we are thinking of harm comparatively, then a child's having a life not worth living doesn't explain how creating that child harms her. If there is something wrong with intentionally creating such a child, then, we must discover the wrongness elsewhere.

We might, then, abandon the language of harm and benefit, and ask instead whether we seem to *have reason* to create or not to create in various circumstances. So let us return to the case of the child whose life we called 'not worth living'—and let's call such a child the 'miserable child' instead. Most people seem to have the strong intuition that we have very good reason not to create the miserable child, if we can avoid it. That is: if you were to find out that getting pregnant right now would result in the creation of the miserable child, you would likely take that to be decisive reason not to get pregnant right now. The serious badness of such a life seems to generate a reason not to create it.

The idea that one has reasons not to create certain children does not seem limited to the case of *miserable* children, though. Consider: if you found out that, by getting pregnant right now, you would create a child who will have very serious, painful, medical complications, doesn't this seem to provide you with a reason not to get pregnant right now? This seems true even if such a child's life, though filled with medical hardships, would be overall worth living. That is to say: there seems to be

---

advice is surprisingly difficult to justify. By changing the time of conception, one changes the identity of the child. And so long as the first child would have a life worth living, then creating that child would not harm it. So it becomes surprisingly difficult to justify the intuitive claim that, say, a 15 year old girl ought to wait until she is older to get pregnant, as doing so 'would be better for the child'.

reason not to create children who will suffer in various ways, even if they will not be 'miserable'.

In response, it might be thought that, while true that the suffering of a potential child provides that child's parents with a reason not to create it, the happiness of the potential child provides countervailing reasons *to* create it. Thus, for a child who will not be miserable, there may yet be, on balance, reason to create it. However, this second claim seems false. As I sit here, it is true of me, given facts about my life, my health, and my environment, that a child I have is likely not to be miserable, and indeed, even to have a life with significant joy in it. Do I thereby, right now, have a reason to make a baby? The answer seems to be obviously 'no'—for all of us, who throughout our lives could be making new people with relatively happy lives, this fact seems to provide *no reason* to go about making those people. While adding happiness to the world by, say, making a sad person happy (perhaps by feeding her, or providing her with medicine) seems obviously good, and something that I obviously have reason to do, it does not seem true that I have a reason to add happiness to the world by *adding happy people* to it.

The result of this analysis of reasons generates the very strange, but very intuitive claim that philosopher Jeff McMahan (1981) has called **The Asymmetry**, which is the following:

The Asymmetry: Although the prospect of pain and suffering in the life of a child provides one with reason not to create that child, the prospect of happiness in the life of a child provides one with no reason to create that child.

The Asymmetry is theoretically very strange, and for this reason has been rejected by some (including McMahan, who named it—see his more recent (2009)). However, it is so intuitive that most people find it hard to deny. Indeed, philosopher Melinda Roberts observes that even those who would like to reject it, based on its theoretical strangeness, feel compelled to accept it, based on its intuitive appeal (Roberts, 2011, p. 336). For rejecting the asymmetry would require accepting either that there is no reason not to create the child who will suffer, or that there is reason to create every child who would be reasonably happy. And both of these options seem unacceptable.

If **The Asymmetry** is true, however, then it has deep implications. Every person or couple considering whether to create a child should consider the interests of their prospective offspring, but it appears that the reasons generated by such a consideration will always be lopsided: each of us will always have *some* reason not to create a child, as that child will experience at least *some* suffering in its life, and no one will have any reason to create a child *for the child's sake*.[7]

---

[7] The strongest argument to be made on the basis of this sort of consideration would look like that of the philosopher David Benatar, who holds that, for each of us, it would be better never to have been born, and so we are each obligated not to impose life on anyone else. The morally best world, then, is one in which the human species goes extinct (Benatar, 2006). Although there is much philosophically interesting to discuss in this proposal, it has convinced very few philosophers. For present purposes, we may simply note it as a 'book-end'—the most radical position that one could take on the basis of some sort of procreative asymmetry; but I, instead, will suggest that one need

But we might think that such considerations, in the best situations, will be relatively unimportant. If my parents, for instance, took such considerations seriously, they may have admitted that my interests generated some reason not to create me (and no reason to create me), but that these reasons were relatively weak. After all, they were able to predict (accurately, it turns out) that I would have a pretty good life, and so the suffering that I do endure does not provide anything like the strength of reason that one has to avoid creating the miserable child. In such a case, we might think that the parents' own interest is of sufficient weight to make procreating permissible. My parents wanted to have a child badly enough that these desires outweighed the relatively weak reasons generated by the suffering that my life would contain. Such a story does not seem implausible.

Unfortunately, our situation today is not like my parents' situation. While my parents had every reason to believe that my life would be even better than theirs, the threats of climate change and overpopulation make such a belief much more questionable today. If my child lives 80 years, then it is overwhelmingly likely that she will live to see and experience food and water shortages, increased disease outbreaks, an increase in deadly heat waves, frequent, catastrophic storms, massive migration by climate refugees, and the economic and political destabilization that is likely to result from these eventualities. Now, even though I can predict that my daughter will be better protected from these harms than the world's poor, she is unlikely to be completely protected from all of them. And the calculation will only get worse with each new generation, until we set ourselves on a sustainable path. Do the interests of my potential children give me reason not to create them, then? Do yours? What about the generation that will bear children in, say, 20 years? As the global situation gets worse, it will become more difficult to believe that the reasons not to have a child—grounded in that child's interests—are negligible. It is harder for me than it was for my parents to believe that my own parenting desires outweigh the reasons generated by the suffering my child will endure. And as this belief gets less reasonable, it will seem more likely that the duty to protect my child will require sacrificing one's own parenting interests for the sake of not exposing a child to the suffering that comes with existing in our world.

3.4 What Might Our Obligation Be?

If any of the candidate moral principles articulated above are plausible, they would seem to have some relevance for our procreative behaviors. Having a large, wealthy, carbon-expensive family would seem to contribute (relatively largely) to a massive systematic harm, to be unfair, and to place many children at risk of serious harm simply by being in the world. Is that all that can be said, though? Vaguely, that there is some moral concern with having a 'large' family? More pointedly: do the moral

---

not accept such a contentious view in order to be driven towards the conclusion that we each have procreation-limiting obligations.

principles articulated in this chapter entail that each of us has a duty to have *no* children? Or perhaps just one or two? In what follows, I will not argue for having some specific number of children, but will articulate the reasons for thinking that we ought to limit ourselves to zero, one, or two children per couple. On the basis of these reasons, I will conclude that, whichever precise number may be correct, it seems plausible that the principles articulated here entail a duty for many of us to have *at most two* children.

Let's begin with the putative duty to protect one's child, as it may seem the most demanding of the candidate obligations. Consider: if each of us is obligated to protect our children by not putting them into a predictably dangerous environment, then the normative implication seems clear: each of us is obligated not to procreate, *period*. Numbers don't seem to matter a lot here—the obligation isn't in terms of minimizing the number of new children exposed to risk; rather, the duty is to *protect one's children*. Each potential child, then, is owed protection in the same way, and it doesn't make sense to say, "Well, I didn't expose *too many* children to harm."

If we have a duty to protect our children (even from existence), and if the world that such children will live in is likely to be sufficiently dangerous to trigger such a duty, then it would seem to be the case that each of us is obligated to refrain from procreating altogether. This would be intensely demanding. But we should note several things at this point.

Although the duty to protect one's children is not, *per se*, attentive to numbers, that doesn't mean that numbers will be completely irrelevant. In the following chapter, we will explore the kinds of considerations that could defeat a *prima facie* duty, and it will be argued that it may matter whether refraining from having any children at all is too costly in various ways. In this way, we might think that interests other than the potential child's can determine the precise content of the duty.

In addition, we can note the uncertainty about the future, and the relevant differences in our predictions at various times. I suggested above that my parents likely had no reason to think they were obligated to protect me from existence, but that I may have *some* reason to think that about my own (potential) children. But these reasons may yet be uncertain or relatively weak, compared to what the next several generations will face.

Finally, it is worth noting that just because 'protecting one's children' is a duty that applies to each child, that doesn't mean that one can't do better or worse by this duty. If a child being exposed to danger is bad, then presumably two children being exposed to danger is worse, and three is worse still. While this doesn't mean that one can discharge her duty by 'protecting most children', it does mean that there are additional reasons not to expose *more children* to danger. If we combine this thought with those above, then, we might think that the obligation of a parent today is to take seriously the future risk and to expose no more children to it than is necessary in pursuing one's own interest.

I'm not sure how successful we should take each of these considerations to be in pushing towards some particular number of permissible children. On the one hand, the duty to protect one's children seems to imply a very demanding obligation: that each of us, if the future is bleak enough, refrain from exposing any new child to our world. However, we might think that, at least for now, the future is sufficiently uncertain that parental desires could outweigh the risk to our children (especially if we have greater-than-average means to protect them throughout life). And in this case, perhaps it is permissible to have a child, even in light of the coming danger. But could it be permissible to have more than one? The case gets more difficult. We will discuss, in the following chapter, how a parent's projects and desires might play a role in justifying the choice to have some number of children; but for now, we can simply note that each procreative choice is a choice to expose *another* child to danger, and so the justificatory bar is raised. The duty to protect one's children, then, plausibly demands that we refrain from procreating altogether, and it at least pushes us to have few children.

The remaining two candidate principles raise the question of 'how many children is permissible' more directly. Consider first the issue of justice from earlier. According to such considerations, it may look like procreating at all would be to commit an injustice, as each new person contributes unequally to a problem that disproportionately harms the poor, and depends on radical inequality. So perhaps justice considerations entail that we have a strict moral obligations not to procreate; this would be an extreme view, but it also looks disconcertingly defensible.

Perhaps, though, we could take a page from Broome's playbook, and say that justice simply requires one to be a 'net-zero' procreator. Because a strict obligation not to procreate would be so demanding, perhaps we could allow the global wealthy to procreate provided that they offset their procreative behavior somehow. This could be through reducing other emissions activities, for instance; or, as Broome suggested in the non-procreative context: one could purchase carbon offsets. Of course, offsetting one's carbon is not a perfect solution, for reasons that have already been mentioned: Offsetting carbon that was liberated from fossil fuels replaces long-term carbon sinks with short-term sinks. In this way, it may be misleading to call emission + offsetting 'net-zero', and so likely could not justify giving the world's wealthy a *carte blanche* concerning procreation. However, procreation + offsetting would be much better than simply procreating, and so perhaps it would allow wealthy individuals to prevent the existence of their children from contributing (as much) to the problem, while not requiring so severe a procreative restriction.

As with the duty to protect one's children, the goal here is not to settle whether justice requires that one *not procreate*, or whether we might be limited to relatively few children; the goal, rather, is simply to demonstrate that considerations of justice seem to imply *some moral restriction* on one's procreative behaviors; it appears that justice points in favor of *limiting* one's procreative activities.

Finally, consider the duty not to contribute to massive, systematic harms. Overpopulation is causing massive, systematic harms, and so we have a duty not to contribute to overpopulation. But what does it mean to contribute to overpopulation? Well, it's unclear. On the one hand, by procreating at all, one becomes a pro-creator—someone who has engaged in the activity that is ultimately causing the problem. So maybe our duty is not to procreate. This, again, would be an extreme view. But there are other ways to understand what it would be to contribute to overpopulation.

One might, for instance, think that procreating past 'replacement' would contribute to the problem, since the issue is not making people, but making people at a rate that grows the population. So, perhaps our obligation is to have no more than two children per couple. Unfortunately since our current population is already unsustainable (especially given the radical inequality discussed above, and the incredible resource-expense of the global wealthy's children), each couple having two children might actually still constitute contributing to overpopulation. So perhaps each couple must have no more than one child, since having one child is compatible with reducing the population to a sustainable level.

Each of the positions laid out above seems at least initially plausible, and I'm unsure on what grounds we might take one of them to be obviously correct. What does seem clear, however, is that having any more than two children would be 'contributing to overpopulation'. And so, if there is a duty not to contribute to massive, systematic harms, then it seems plausible that we each have an individual obligation to limit our own procreative behaviors to the creation of two new children per couple.

Before closing out the chapter, I want to revisit an issue from Chap. 2—that of the difference that one can make towards the harms of climate change thought of in *relative* rather than *absolute* terms. The motivation for exploring the moral principles in this chapter was the admission that having a child may not make an absolutely significant difference to the harms of climate change, and so we needed to see whether **Significant Difference** is true. If the principles investigated here are plausible, then it is not—we sometimes have obligations to refrain from contributing to a moral problem, even if our contribution doesn't make a significant difference.

Now recall that procreating does seem to make a *relatively* significant difference to the harms of climate change, in that it has a larger impact than anything else we are likely to do in our lives. What is important to see now is that this relative significance turns out to matter, if we have an obligation not to contribute to massive, systematic harms. Why? Because not all obligations are equal, and it's overwhelmingly plausible that the relative contribution to the problem *does matter* when thinking about how seriously we take such an obligation.

In motivating this chapter's non-consequentialist principles, I used examples like recycling, or wasteful driving, in which one's actions are both absolutely and relatively insignificant. And we may in fact think that there is a duty to recycle and not to engage in wasteful driving. However, we might also not take those duties very seriously, as a result of how little they contribute to the relevant moral problems. If taking my motorcycle to the race track gives me a lot of pleasure, then we might shrug off the fact that there is a duty not to contribute to massive, systematic harms, as very many of my other actions could have a much greater effect in the fight against climate change.

The issue here is that, when we specify moral principles into guides for action, there are almost always competing goods: I may decide to eat meat so as not to offend a host, even if morality requires vegetarianism; or I might commute further than would seem justifiable, in order to make my spouse's commute more reasonable. And not all justifications need be high values; our pleasure, joy, and ability to lead flourishing lives matters as well, which is why we might think that the occasional pleasure-cruise is permissible, or even that one could live in a larger house than she strictly requires.

Now certainly, if there is even a *prima facie* duty not to contribute to massive harms, and things like pleasure cruises, large homes, vacations, etc. contribute to climate change, then we would expect that there is a justificatory burden on anyone who wants to do these things. Meeting this burden, then, may require that one not do all of them, or not very often, and that one attempt to minimize the cost (buy a fuel-efficient vehicle, increase the energy efficiency of one's home, etc.). But the primary point here is that if we really have a duty as wide-ranging as the duty not to contribute to massive harms—and if, as is the case, virtually everything we do contributes to the harms of climate change—then it seems plausible that the duty will be sensitive to various features of one's contributions, such as the relative significance, as well as whether one attempts to offset and/or compensate for those contributions.

Clearly, this final point is relevant to our discussion of procreative ethics. What seems plausible in many of the examples of emitting activity given is that the increase in significance of one's contribution as a result of various actions increases the justificatory burden of those actions. If I want to jet-set around the world, doing so would seem to demand much stronger justifications than taking my bike to the track; while the modest joy I get from the latter may do the justificatory trick in that case, the jet-setting would seem to require that I have much better reasons for acting.[8] And of course, if that story sounds plausible, then the decision to have a child is perhaps the most in need of justification of all our potential actions. According to the study of 'carbon legacy' from the previous chapter (Murtaugh & Schlax, 2009), in the near term, having a child may make one responsible for as much as 9441 metric tons of CO_2, while a flight from Washington, DC to Paris increases one's carbon footprint by approximately 1 metric ton. While one ton is nothing to sneeze at (recall that the average Nigerien has an *annual* footprint of 0.1 tons!), having a child is nearly *10,000 times* more costly. If flying across the ocean requires justification (and it seems to), then having a child requires *much, much more*. And this, I take it, is a surprising conclusion.

3.5 Conclusion

A serious puzzle about moral problems like overpopulation and climate change are how they can generate individual obligations, given that one's personal contribution to the problems are vanishingly small. After all, what makes these issues

---

[8] Note that this need not imply that the justificatory burden could not be met. Physicians who work for Doctors Without Borders presumably have a large carbon footprint as a result of their international travel; I would think that their relatively large contribution to climate change is justifiable in the way that flying from New York to Paris for a fancy lunch would not be.

problematic in the first place is that they cause harm; so if my individual actions don't actually cause any of that harm, or make a significant difference to the problem, how could they be wrong?

In this chapter, I have argued that, intuitive though such an objection is, it relies on too simplistic a view of what grounds individual moral obligation. Although I have not argued that any particular moral theory is correct, I have tried to show that according to three, plausible moral principles, we inherit moral obligation from the problems of overpopulation and climate change. In short, it seems plausible that we have duties (a) not to contribute to massive, systematic harms; (b) not to act unjustly; and (c) not to have children who will have bad lives. If any subset of these arguments seems convincing, then each of us likely has an obligation to limit our procreative behaviors, even if our individual actions don't make a significant difference. And while it is unclear whether there is some particular number of children the creation of which is permissible, the clear justificatory push is for *small families*. Indeed, considerations of justice and protection of one's own children push in the direction of having *no* children, while the duty not to contribute to harms suggests a limit of two children per couple.

The candidate moral principles here investigated, then, suggest that each of us is obligated *at least* not to procreate past replacement, with some non-trivial chance that the moral burden is even stricter than that. If true, such a conclusion is morally profound, and to many, likely disconcerting. It will seem to many that there must be good justification for having children, which seems so valuable. And indeed, we have just seen that the existence of a relevant duty does not mean that one *must absolutely do* what is required; but rather, that one must do that thing unless she meets the appropriate justificatory burden. However, we have also seen that, due to the relative contribution of procreating to the problem of climate change, the justificatory burden is very high—likely higher than for any other single activity that most of us will take in our lives. Could any reason for procreating meet that justificatory burden? It is to precisely this question that we now turn.

References

Benatar, D. (2006). *Better never to have been: The harm of coming into existence*. Oxford: Oxford University Press.
Broome, J. (2012). *Climate matters: Ethics in a warming world*. New York: W.W. Norton.
Fruh, K., & Hedahl, M. (2013). Coping with climate change: What justice demands of surfers, mormons, and the rest of us. *Ethics, Policy and Environment, 16*(3), 273–296.
Hedahl, M., Fruh, K., & Whitlow, L. (2016). Climate mitigation. In B. Hale & A. Light (Eds.), *Routledge companion to environmental ethics*. New York: Routledge.
Intergovernmental Panel on Climate Change. (2014). *Climate change 2014: Impacts, adaptation, and vulnerability*. Cambridge: Cambridge University Press.
Jamieson, D. (2014). *Reason in a dark time: why the struggle against climate change failed—and what it means for our future*. Oxford: Oxford University Press.
McMahan, J. (1981). Problems of population theory. *Ethics, 92*, 96–127.

McMahan, J. (2009). Asymmetries in the morality of causing people to exist. In M. A. Roberts & D. T. Wasserman (Eds.), *Harming future persons: Ethics, genetics and the nonidentity problem* (pp. 49–68). Dordrecht: Springer.

Murtaugh, P. A., & Schlax, M. G. (2009). Reproduction and the carbon legacies of individuals. *Global Environmental Change, 19*, 14–20.

National Oceanic and Atmospheric Administration. (n.d.). *What is ocean acidification?* Retrieved January 29, 2016, from pmel.noaa.gov: http://www.pmel.noaa.gov/co2/story/What+is+Ocean +Acidification%3F.

Norcross, A. (2004). Puppies, pigs, and people: Eating meat and marginal cases. *Nous-Supplement: Philosophical Perspectives, 18*, 229–245.

Parfit, D. (1984). *Reasons and persons*. Oxford: Oxford University Press.

Roberts, M. A. (2011). The asymmetry: A solution. *Theoria, 77*(4), 333–367.

Singer, P. (2010, June 16). Last generation? A response. *The New York Times*. Retrieved from http://opinionator.blogs.nytimes.com/2010/06/16/last-generation-a-response/.

Singer, P. (2010, June 6). Should this be the last generation? *The New York Times*. Retrieved from http://opinionator.blogs.nytimes.com/2010/06/06/should-this-be-the-last-generation/?_r=0.

The World Bank. (2011–2015). *CO2 emissions (metric tons per capita)*. Retrieved January 29, 2016, from data.worldbank.org: http://data.worldbank.org/indicator/EN.ATM.CO2E.PC/count ries/1W?display=default.

United Nations, Department of Economic and Social Affairs, Population Division. (2015). *World Fertility Patterns 2015—Data Booklet (ST/ESA/SER.A/370)*.

Chapter 4
Challenges to Procreative Obligation

If the arguments of this book are more or less on track, then it looks as though each of us may have an obligation to make procreative choices restricted by concerns about overpopulation and climate change. In particular, it looks as though each of us may have an obligation to have no children, only one child, or at most two children, depending on what we think about the defensibility of the various principles investigated. When faced with such a claim, however, many people have a hard time believing it. Could morality really invade so far into the private sphere? Could it be so demanding as to require that we give up having the size of family that we want?

This reaction is understandable, and there is significant philosophical literature on what morality can demand from us, as well as whether it could intrude into our procreative decisions, in particular. These issues will be the topic of this chapter. In line with the general methodology of this book, I will propose what I take to be the most powerful objections to the idea of procreation-limiting obligations, in order to see whether such objections can really get us entirely off the moral hook. And, perhaps unfortunately, I will conclude that they cannot. While the objections to be investigated here provide good reason to think that we are generally not obligated to remain childless, it is much less clear whether such objections are able to defeat a more modest obligation to have no more than one or two children.

4.1 Being Good Can Be Hard

According to the arguments of the previous chapter, having a child contributes to massive, systematic harms and to injustice, and it exposes one's child to a potentially dangerous world. Further, it seems plausible that we have strict obligations not to contribute to massive, systematic harms, not to commit injustice, and not to expose our children to a potentially dangerous world. Thus, it looks disconcertingly

© The Author(s) 2016
T.N. Rieder, *Toward a Small Family Ethic*, SpringerBriefs in Public Health,
DOI 10.1007/978-3-319-33871-2_4

plausible that we could have a strict obligation not to have any children, so as to completely avoid violating any of these principles.

An obligation not to have any children, however, would be very demanding. If such an obligation were actual, it would require that very many people give up a very central desire—that of having children—for the sake of avoiding wrongdoing. Does this constitute an objection to such a suggestion?

Some philosophers have thought that a general concern with demandingness does, in fact, constitute a legitimate objection against arguments like the one I have put forward. Interestingly, this objection is typically raised against utilitarian arguments, as the utilitarian position tends to require fairly radical sacrifice on the part of moral agents,[1] and we have seen that the arguments from Chap. 3 are distinctly *not* utilitarian.[2] Indeed, these arguments are what we uncovered in our attempt to make sense of environmentalist intuitions, even if emitting carbon 'makes no meaningful difference' to the problem of climate change. However, what has come to be known as the 'demandingness objection' is not only applicable to utilitarian arguments, as we are now observing. Indeed, all that matters for a demandingness objection to get off the ground is that one consider a candidate obligation that, given context, seems to imply significant sacrifice on the part of ordinary agents. Utilitarianism (so the charge goes) does this regularly; but if the arguments from Chap. 3 are on target, our non-utilitarian principles also raise demandingness concerns.

So why think that a moral principle's being demanding constitutes an objection to that principle? It is a good question, and the answer isn't clear. One might respond that ethics ought to fit the kinds of beings that we are—imperfect humans, not angels—and that imperfect humans can't be expected to give up so much for morality. But what does this mean? It can't mean that humans are *incapable* of, say, refraining from having children; after all, many people so refrain. In general, it will appear false to say that we are psychologically incapable of attaining some level of moral goodness; it may be *hard* to be good, but it's not impossible.

So could the challenge really just be that morality shouldn't be too hard? That doesn't seem right, either. In all likelihood, morality *will* be challenging. Living a good life, discharging one's obligations and being virtuous are likely to be exceed-

---

[1] See, for instance, the argument made by Peter Singer that each of us is obligated to donate to famine relief up to the point of 'diminishing marginal utility'—that is, up to the point at which each of us would be made worse off than those we are trying to help (Singer, 1972).

[2] For those who need a primer on moral theory: utilitarianism is a form of consequentialism. Whereas consequentialism, in its broadest form, states only that the rightness or wrongness of an act is determined by its consequences, utilitarianism specifies that view a bit. According to utilitarianism (which is typically presented in its *maximizing* form): an act is right if and only if it, among all of an agent's possible actions, best promotes the balance of happiness over unhappiness. Thus, utilitarianism tells us *how* consequences determine the rightness of an act—by maximizing the happiness/unhappiness ratio. Now we can see why it is interesting that demandingness is being raised as an objection to principles articulated in Chap. 3, as those principles were particularly *non-consequentialist*, and demandingness often seems to arise as a challenge to utilitarianism, which is a form of *consequentialism*.

ingly difficult. If we have an obligation to prevent suffering when it's easy, and our world is full of easily-preventable suffering, then discharging this obligation will be a serious challenge; but it's unclear why the level of difficulty should erase the duty.

To be sure, some philosophers do, in fact, think that fairly straightforward considerations of demandingness might undermine particular principles or theories, and we could spend a significant amount of time investigating this debate. However, my few words here are intended to reveal why I am skeptical that the issue is really about demandingness as 'difficulty' or 'challenge'. Instead, I suspect that the most charitable way to read most objections from 'demandingness' is as a concern with what Bernard Williams calls one's *integrity*.

4.2   Maintaining Integrity

In a justly famous essay, Bernard Williams articulates one of the most powerful objections to utilitarian moral theories, which has come to be known as the integrity objection (Williams, 1973). Now, as we have just discussed, the principles under investigation in this book are not utilitarian, and so Williams' objection does not directly apply; however, as was the case with the straightforward demandingness objection, it is easy to see how a worry about one's integrity could apply to non-utilitarian positions as well. For the sake of explication, let us look first at Williams' actual argument, and then see how it applies to the current discussion.

The key idea of Williams' argument is what he calls the 'negative responsibility' demanded by utilitarian theories. Because utilitarianism requires that every agent maximize total happiness, and because the view does not distinguish between promoting happiness by acting or by allowing another to act, each agent finds herself in the position of needing to calculate what *others* will do, given one's actions, and then to act so as to maximize happiness given this fact about everyone else.

To illustrate this point, Williams raises two examples—those of George the chemist and Jim the explorer—each of whom finds himself in a position to make the world better by preventing another from acting. George has the opportunity to take a (much needed) job researching and developing chemical weapons, which he detests, but which will prevent a very enthusiastic colleague from taking the job and pushing research forward at an alarming rate. And Jim finds himself in a situation where, if he is willing to shoot and kill a single person, then the tyrannical sheriff of a small village will refrain from shooting that person, as well as 19 others.

In both cases, utilitarianism holds that the morally correct action is obvious: George should take the job, and Jim should shoot the villager. This is because utilitarianism doesn't distinguish between one's responsibility for the direct, intended consequences of her own action, and the unintended, but foreseeable consequences of her inaction. Since George knows that turning down the job won't help slow weapons research, but will actually hasten it, and Jim knows that refusing to shoot the villager won't prevent his death, but will instead result in his, and 19 others, dying anyway, George and Jim are responsible for the consequences. This is the doctrine of negative responsibility.

The problem with the utilitarian's picture of negative responsibility, Williams argues, is that it threatens to take away an agent's *integrity*. As autonomous, moral agents, we believe that we are permitted to have some personal projects, and that it matters *what we do*. If I'm a pacifist, then I must resist violence, even if violence might occasionally make the world a better place. And if I'm a parent, then I must treat my children with special deference, even if doing so might be sub-optimal from an objective perspective. The problem with utilitarianism, Williams claims, is that it requires that we hold all interests so loosely that, as soon as the world conspires to make pursuing those interests sub-optimal in terms of happiness promotion, we must let them go. But if we are committed to not giving any special consideration to our own interests and projects, then we don't actually *have* any projects. In what sense am I a pacifist if I know that, as soon as violence would be helpful, I will embrace it?

The key insight of Williams' argument here is that utilitarianism requires each person to see herself as a mere cog in a great happiness-producing machine. But such a cog has no protected projects, and so no integrity. As summarized, the argument applies quite specifically to utilitarianism, as the threat to one's integrity comes from the negative responsibility implied by the constant requirement to promote happiness. However, it is common to hear philosophers use the language of an 'integrity objection' against demanding, non-utilitarian theories as well. It is in this spirit that I raised the possibility that there may be an integrity objection against the view that we all have an obligation not to procreate. But if this derives—as it does here—not from a utilitarian argument, what might such an objection mean?

In short, my sense is that Williams' integrity objection has been taken up as broader than he intended as a result of the anti-utilitarian view that moral agents, if they are to retain their integrity, must have some normative protection for some of their personal projects. Although developed by Williams against utilitarianism specifically, the general idea that morality ought not to threaten our integrity might be useful against many views. On this looser understanding of the critique, we might hold that moral agents must be allowed some 'normative protection' for at least certain projects—that is, that we must be allowed to pursue some activities, even if they conflict with plausible moral demands.

Now of course, not just any project should be granted normative protection: there is no protection for the project of becoming the very best torturer, for instance. And so there must be conditions on what might count as a legitimate, candidate project. Minimally, to rule out the torturer, we probably want to restrict candidate projects to those that are morally valuable. In addition, mere desires or whims likely shouldn't be protected from moral demands; although I happen to like the taste of animal flesh, this mild preference would not protect my carnivorous habits if morality turned out to require vegetarianism.

So what might be a plausible candidate for the status of normatively protected project? Well, having children, as it turns out, seems to be a pretty good one. Most of us tend to think that creating a child and raising it in a loving home, with one's own values, is a clearly *valuable* endeavor. Indeed, many see it as one of the greatest goods in life. And our desires—for those of us who have them—to form a family in this way are not 'mere' desires or whims: they tend to be central to our very being, structuring much of the rest of our lives. Indeed, many parents report having children as being one of the activities that gives meaning to their lives.

If this suggestion is plausible, then the seeming demandingness of an obligation not to procreate is more charitably thought of as a threat to our integrity as agents. And this concern with our 'integrity' reveals a belief that we ought to be granted normative protection for at least some of our valuable projects—that is, that the pursuit of an otherwise good project can sometimes be morally allowed, even in the face of plausible moral demands that one refrain. And since creating and rearing a child in a loving home is a central, meaning-giving project to many prospective parents, then a concern with integrity looks plausibly to support the view that procreation deserves 'normative protection' from the demands of duty.

So what, precisely, does this mean for our candidate claim of obligation? It seems to imply that, at least for those with a procreative project, there is no strict duty or obligation to remain childless; that is: having a child is permissible. But does a concern with integrity imply that having more than one child is permissible? Might there still be an obligation to have no more than one child?

Interestingly, this last possibility does not seem ruled out by the integrity objection. Plausibly, the integrity objection is so powerful against the suggestion that there is an obligation not to have any children because it requires that each person *never procreate*. And so we can imagine a person—likely many of us personally know someone like this—for whom creating and raising a child is a central life project. Such a person might say that reproducing gives meaning to her life, or that, were she not allowed to pursue reproduction, her life would be seriously lacking. However, after someone has had one child, her situation is importantly different. A requirement not to have another child *is not* a requirement to live forever without having had the opportunity to create a child; rather, it is the requirement that she have *no more* children. Does such an obligation plausibly threaten one's integrity?

It's unclear that it does. What makes the integrity objection so compelling in the original procreative case is that an obligation not to have children would require that one never experience a unique—and uniquely valuable—human good. And indeed, if a prospective parent had no ability to adopt an already existing child, the obligation would also require such a person to remain forever childless.[3] If any human projects deserve normative protection from moral demands, we must include going through the creation process and starting a family in that class. However, once a parent has a child, requiring that she not procreate again does not deny her either of

---

[3] I do not mean to ignore the possibility of adoption—indeed, I address that question specifically in my (Rieder, 2015). If one wants only to *parent*, and not necessarily to *procreate*, then a parenting project could justify only adopting a child, rather than making a new one. Of course, as I argue in (2015), some people also have a particularly procreative project—in particular, some women have a *gestational* project—and this seems also a good candidate for generating normative protection from otherwise plausible duties.

those goods; rather, she will simply be required to maintain her family's current size. And having a large family—or a family of any particular size—does not seem either uniquely valuable, or like the sort of feature on which life's meaning should depend. While being forced to remain childless, or to miss out on human creation, might justifiably be seen as a tragedy, being forced to have an only child does not seem like a tragedy. At most, it seems unfortunate, or perhaps like the thwarting of one's desires.[4]

4.3 Procreative Liberty

In the previous sections, I have argued that demandingness objections—if taken to mean simply that morality ought not to be too hard—are unconvincing, but that interpreting demandingness as a concern about integrity meets with more success. Plausibly, protecting the integrity of moral agents rules out a strict moral obligation not to procreate. However, the slightly weaker suggestion that one ought not to have more than one child does not seem a threat to one's integrity, and so it may yet be the case that each of us has a duty to have no more than one child.

This weaker candidate obligation will still not sit well with many people. To some, it will seem that *any* obligation not to procreate is implausible, as people have the right to decide whether and when they will have children. And indeed, this idea has lots of support in philosophy and in law, under the name 'procreative liberty' or 'procreative freedom'. In this section, then, we will consider briefly whether a concern for procreative liberty might save us from an obligation to have no more than one child.

The United Nations has articulated well the widely-embraced idea of procreative liberty, declaring in a conference on human rights that "[p]arents have a basic human right to determine freely and responsibly the number and the spacing of

---

[4]It has been suggested to me that there are at least two other candidate projects that could justify having more than one child: the project of having a 'big family', and the project of having both a boy and a girl. Although I admittedly lack the space here to thoroughly analyze such prospects, I will simply note that I am unconvinced that these rise to the level of projects, such that they should be given protection. Both seem like cases of potential desires—perhaps even very strong desires— but they seem different in kind from the *project* of having a child at all. In going from being childless to having a child, one goes from being a non-parent to being a parent. This fundamentally changes who one is. It is hard to imagine that the addition of more children can have a similarly profound effect, and it's partially this *uniqueness* that seems plausibly to justify having that first child. We might have further concerns about the project of having both a boy and a girl, grounded as it is in biological sex. We might think that caring so much about the differences between raising a little boy and a little girl are in fact inappropriate.

It is possible, of course, that I'm wrong, and that these (and potentially other) projects would ground normative exemptions from procreation-limiting obligations. Even in that case, procreation past the first child would require justification, and only those with legitimate projects could do so permissibly. My thanks to Marcus Hedahl for suggesting that I consider these other possible projects.

their children" (United Nations, 1968, p. 4). In addition, moral philosophers some-times advocate allowing parents the freedom to create children *how* they want on the basis of procreative liberty considerations; that is, procreative liberty not only grants parents the right to choose the number and spacing of their children, but also whether to utilize legal reproductive technologies. Jonathan Glover, for instance, argues that the general deference owed to procreative liberty justifies considerable discretion for parents to choose their children through preimplanta-tion genetic diagnosis, genetic testing plus abortion, or even genetic manipulation (Glover, 2006).

When invoking the language of 'procreative rights', we need to be very careful about what, precisely, we are discussing. The hurried move from the existence of 'procreative rights' to a blank permission slip slides over multiple, difficult issues. The first of these seems fairly elementary, but can trip up even a careful reasoner, and that is the distinction between legal rights and moral rights. The difficulty here is that we tend to think that both of these types of normative protections exist—and that they are likely related to one another—but that they are not identi-cal. A legal right owes its existence to a system of rules or laws, or the decisions of an authoritative deliberative body, and it grants permissions from and protec-tions by the state to individuals. If I, as an American, say that I have a right to due process, I am most likely invoking my *legal right*, guaranteed by the Constitution. I could suggest that *morality* demands due process, but that would be a far more abstract claim then I need to make. By virtue of being an American citizen, I can invoke my right to due process as a protection against certain treatment by agents of the government, and this implies that such agents are legally obligated to pro-vide me with due process.

One thing, then, that one might say in response to the proposal that we have procreation-limiting obligations, is that each of us has *legal* procreative rights. And indeed, this seems true in many countries—certainly in America. We think that it is not the government's business how many children we have, or how quickly we have them, and this belief justifies many people's resentment of China-style family planning policies. Further, we expect that, were concerned citizens to try to force any of us to change our procreative behaviors, the state would step in and use force to protect us from the interference of others. The invoking of legal rights in the procreative context conveys the idea that procre-ation is an essentially private act, and a belief that individuals should be permitted by the state to act as they wish.

Legal rights, however, are not the issue under investigation here. If procreative rights are to protect us from procreation-limiting obligations, then they must be more than legal rights. This is because we are obligated to do many things that aren't required by law. Consider the question: if you have promised your best friend that you will help him move today, and you don't feel like it, do you have the right to break your promise? Well, it's certainly your *legal* right to break such a promise; you will not be arrested or fined, and no one will physically compel you to keep your promise. But this is a paradigmatic case of *not* having a moral right. Rather, you have a duty to help your friend, and your friend has a right against you that you come and help him. To have a moral right, even in the absence of a legal right, is to have a moral *permission*. It is thus the having of *moral* procreative rights that would save us from procreation-limiting obligations. And so we can specify the question of this section helpfully: do we have, in the words of the Proclamation of Tehran, a *moral right* "to determine freely and responsibly the number and the spacing of their children"?

This question is much more difficult. It certainly seems true that we have some degree of moral rights concerning our procreative behaviors. However, the language of 'some degree' is crucial. We tend to think that virtually no liberties are absolute; after all, although you are generally free to swing your arms in the air, this freedom is contingent on my face not being in the path of your hand. We recognize restrictions on our freedoms all the time—typically for the sake of promoting others' interests. Dan Brock, for instance, suggests that if a couple is living in a very resource-poor area, it may not be permissible to have, say, a third child, when doing so threatens the entire community's access to goods. In such a case, not only does one's procreative liberty seem restricted by the interests of others, but those others may well even be justified in demanding that one not have the third child (Brock, 2005).

Does Brock's case apply to our own situation? It's simply not obvious how strong the moral case against one's action must be in order to undermine one's moral rights. I tend to think that our rights evaporate fairly quickly when they come into contact with the interests of the larger population. And for that reason, I (and others who think similarly) are relatively comfortable with the idea that considerations of, say, *public health* can obligate me to act contrary to my desires. By analogy, we can think of the case of vaccinating one's child: while there are those in American society who desire not to vaccinate their children against dangerous diseases, we might argue that the interests of others (and the population as a whole) can undermine one's otherwise stable right to raise his child as he sees fit. If such a claim seems plausible, then our case against unlimited procreative rights gets stronger.

Determining whether moral rights can stand up to concerns about the population's interests would take at least another book, and so I cannot provide a convincing argument here.[5] So instead, I will note only that the more one tends to think that a populations' interests can undermine individual moral rights, the more likely one is to think that Brock's case extends to our current situation. But since I cannot here provide a compelling argument, I will instead concede for the sake of this project that our moral rights may not be limited by others' interests so strictly as I tend to think they are, and so it may be the case that each of us has a moral right to determine freely the number and spacing of our children.

---

[5] As I finalize this manuscript in early 2016, philosopher Sarah Conly has released precisely such a book, defending the idea that we do not have unlimited procreative rights. In particular, she argues that considerations like those raised here have the implication that families have a right to only one child (of course, this is a right of non-interference, not a right to be given a child). While the argument I am making here will end up being significantly weaker than Conly's, those who are interested to see a full defense of the strong one-child view should consult (Conly, 2016).

4.4 Rights, What Is Right, and the Right to Do Wrong

If each of us has fairly unrestricted procreative moral rights, then one last question remains: are we thereby off the moral hook? Perhaps unfortunately, I think the answer is 'no'. The argument for this claim will proceed in two steps. In this final section, I will make the fairly simple point that acting within one's rights does not entail acting *rightly*, and that in fact, we might think that there is even such a thing as the *right to do wrong*. Making such a case, however, requires that there be other moral theoretic tools that could tell us what morality recommends, and investigating some of these tools will be left as the task of Chap. 5.

Suppose, as many people would likely find plausible, that each of us has a moral right to our earned, accumulated wealth. The question of interest now is then: must morality therefore be silent concerning what I do with my wealth? If it seems at least appealing that morality must be silent, consider two cases: in one, I spend every last dime on a life of wanton hedonism; and in the other, I spend a significant percentage of my fortune on alleviating suffering in the world, and the remainder on providing for my family. Are these two options morally equal?

Most people feel compelled to admit that these cases are not morally equal, even though I act within my rights in both cases. And that's because acting within one's rights does not entail acting *rightly*. If I ask what I morally ought to do with my wealth, there seems to be correct and incorrect answers, even if we agree that all of the options are within my rights.

It is confusing that this might be so, as we tend to think of morality is fairly binary: there is right and wrong, good and bad. But the moral concepts that we invoke are not simple. Having a moral right grants us some kind of moral protection, but does not seem to convey that a particular act is *morally recommended*, or what one *ought to do*. In short: having a right to act doesn't imply that it is right to act. So what, precisely, does the protection of a moral right amount to?

Well, recall that procreative rights were invoked as a defense against the claim that there may be an obligation or duty to limit one's procreative behaviors. And this does seem to be implied by the having of a right: if I have a right to my wealth, then I have no duty or obligation to do anything in particular with my wealth. This concept of a duty or an obligation (I'm using them interchangeably) is taken to be especially invasive, as they generate moral *requirements*. If I have a right to my wealth, then I am not required to do anything in particular with it. Additionally, some philosophers think that such a right entails that no one has a *claim* on my personal wealth, and that no one would have the *standing to demand* that I do anything in particular with my wealth. Having a right, then, protects us from the strictest, most second-personally invasive aspects of morality.[6] But it is perfectly coherent that one could be free from obligation to do something, and that doing that thing is still morally good, morally better, what morality recommends, or even the right thing to do.

In fact, some philosophers go so far as to say that one can act *wrongly* while acting within her rights. Indeed, Christine Overall thinks that this is precisely what is

---

[6] For a full treatment of duties and rights as richly second-personal in this way, see (Darwall, 2006).

often going on in the procreative context—that although we have a moral right to create new people, doing so is often wrong. What we have in this case is a 'moral right to do moral wrong' (Overall, 2012).

Maggie Little makes a similar argument for several cases in which profound intimacy seems to ground a protection from the invasion of obligation, but in which it still seems possible to act wrongly (Little, 2005). She offers sex and marriage as instructive act-types that are not typically the object of a positive obligation, but the withholding of which can nevertheless be wrong. Intuitively, we might think this is because one always has the right to choose not to have sex, and one has the right not to get married. However, if having sex with someone would do a great amount of good, and there are no good reasons not to, and yet one withholds sex because the potential partner is Black, then plausibly she acts wrongly. And if a man turns down a marriage proposal that would make him, the suitor, and very many others very happy, and when there is no good reason to refuse, but because the suitor does not have large enough breasts, then this act, too, is plausibly wrong. In both cases, the individuals were within their rights to act as they did; but we might think that they were wrong to exercise their rights.

These cases point to situations in which one has a right to perform some act that there is no good reason to perform, and very good reason not to perform. If the only thing to be said in favor of refusing sex is that one doesn't want to have sex with a Black person, then there seems to be a serious moral problem with the refusal. However, it also doesn't seem to be the case that having only such an ugly, racist reason takes away someone's right: we still don't get to *demand* that the person have sex, as sex isn't the kind of thing that the moral community has the standing to demand.

And so it is, I think, in the case of procreation. We do have procreative rights, and those rights plausibly include the right to decide the number and spacing of one's children; however, such a right does not imply that any particular procreative act is good or right. Refraining from procreation may be beyond the scope of what we, as a moral community, have the standing to demand of someone, but this does not eliminate the very good reasons against having a child. And so procreating too much, or with too little thought, or because one was too lazy to avoid it, may still be morally criticizable. If moral reality is messy in the way I have suggested, these acts may be wrongs that we have the right to perform.

4.5 Conclusion

This chapter may have seemed a bit unsatisfying. I have not offered definitive arguments for one view of our procreative obligations over another, but rather have offered some sketches of how such arguments would have to go. But this doesn't mean that we haven't made any progress. So let's pause for a moment to take stock.

According to the arguments of this chapter, it is implausible that there is a strict moral obligation not to have any children. This is not because such an obligation would be too 'demanding'—understood in the sense that it would make morality too hard to follow—but rather, because doing so would threaten our integrity as agents. The same is not obviously true, however, for the claim that we have a strict obligation to have no more than *one* child. It is consistent with maintaining our integrity as moral agents that we are all morally required to have no more than one child.

The issue of procreative freedom, however, is more complicated. If it is true—what so many seem to believe—that we have the moral right to decide the number, spacing, and method of creating our children, then perhaps it is permissible to have *any* number of children. Whether or not this is so depends on whether procreative liberties are appropriately limited by the interests of others. Unfortunately (for the goal of coming to a clear conclusion!), it is not obvious in either case whether we ought to think of procreative freedoms as limited. Thus, it may be the case that invoking procreative freedom is the silver bullet against the arguments of Chap. 3, establishing definitively that we have no obligation to refrain from procreating.

Although I happen to find it plausible that the interests of others do, in fact, limit our procreative freedom, I have admitted that I cannot make that case definitively (at least not in this book). And so I will, instead, ask what we should conclude about our individual moral burden, assuming that we do not have any obligations to restrict our procreative activities. This is important, I think, because philosophers often act as though establishing a *right* to act ends the conversation. But this isn't true. Even if we each have an unlimited right to procreate, there is much more that we can say about particular procreative acts that is morally relevant. Surprisingly to some, it may be the case that our right to procreate is a 'right to do wrong', or at least that refraining from procreation is what morality recommends, or what we ought to do.

Of course, if any of those claims were true, we would need an explanation for what makes them so. It couldn't be because we are obligated to refrain from procreating, as I have conceded that we may not be. And so, I will spend the final chapter investigating what other moral concepts may have to say about our procreative behaviors.

References

Brock, D. W. (2005). Shaping future children: Parental rights and societal interests. *Journal of Political Philosophy, 13*, 377–398.

Conly, S. (2016). *One child: Do we have a right to more?* Oxford: Oxford University Press.

Darwall, S. (2006). *The second person standpoint: Morality, respect, and accountability.* Cambridge, MA: Harvard University Press.

Glover, J. (2006). *Choosing children: Genes, disability, and design.* Oxford: Oxford University Press.

Little, M. O. (2005). The permissibility of abortion. In A. Cohen & C. Wellman (Eds.), *Contemporary debates in applied ethics.* Oxford: Blackwell.

Overall, C. (2012). *Why have children? The ethical debate*. Cambridge: MIT Press.

Rieder, T. N. (2015). Procreation, adoption, and the contours of obligation. *Journal of Applied Philosophy, 32*(3), 293–309.

Singer, P. (1972). Famine, affluence, and morality. *Philosophy and Public Affairs, 1*(1), 229–243.

United Nations. (1968). *Final Act of the International Conference on Human Rights: Teheran, 22 April to 13 May 1968*. New York: United Nations Publication. Retrieved from http://legal.un.org/avl/pdf/ha/fatchr/Final_Act_of_TehranConf.pdf.

Williams, B. (1973). Integrity. In J. Smart & B. Williams (Eds.), *Utilitarianism: For and against* (pp. 108–117). Cambridge: Cambridge University Press.

Chapter 5
Toward a Small Family Ethic

The goal of this final chapter is to fill out a richer picture of the morality of procreation than one that is exhausted by the concepts of duty, obligation and permissibility. According to the slow, steady retreat of this book, it may be the case that—although there was a compelling argument for various procreation-limiting obligations—perhaps none of them actually stick. That is: perhaps it is the case that we are not *obligated* to have any particular number of children.

Even if this is the case, it is important to realize that we are not necessarily off the moral hook. Although people sometimes talk as though establishing a moral right to act, or establishing the lack of a duty to refrain from acting, entails that the action in question has passed all relevant moral tests, this is an impoverished picture of morality. Some of this has already been indicated in the previous chapter with the cases of having a right not entailing that an action is 'right', and even with the cases of having a 'moral right to do moral wrong'. So now we are left with the question: what is it that we can say about actions that do not violate a duty?

In fact, I believe that we can say quite a lot—and indeed, that we regularly do say quite a lot. Permissible (or even obligatory) actions are often done by mean, selfish, bigoted, cowardly, or otherwise vicious people, and it has long been recognized that these character judgments can come apart from judgments of permissibility. Further, *moral reasons*, even when they don't add up to obligation, can be very important to our moral evaluation of the situation. If the moral reasons clearly favor a certain act, then even if that act is not obligated (or if refraining from acting is within the actor's rights), we might say that the act is what morality *recommends*, or what one *ought* to do. Finally, the way in which an agent responds to the reasons available may tell us much about her, and we may blame or praise her for her actions as a result—regardless of the permissibility of the act.

In the space that remains of this book, then, I will expand on each of these moral dimensions—those of character, reasons and blame and praise. In short, I will argue that including these moral features in our analysis makes for a more plausible picture of the moral phenomena, and reveals that there is significant room for moral judgment outside of considerations of duty, obligation and permissibility. In the

© The Author(s) 2016
T.N. Rieder, *Toward a Small Family Ethic*, SpringerBriefs in Public Health,
DOI 10.1007/978-3-319-33871-2_5

procreative context, then, it appears that this is uncomfortably relevant, as the decision to have many children (or children at all?) looks potentially able to reveal that one has bad character, that one is insensitive to reasons, and sometimes that one is blameworthy as a result. Morality, it appears, may still have the tools to condemn us for our tendency to have babies—even if we have the right to do so.

5.1   (Green) Virtues

Recall the opening challenge to climate ethics from Chap. 2—the suggestion that, since each of our own, miniscule contributions to the climate crisis don't make a significant difference, there can't be an obligation not to make them. Although this principle of **Significant Difference** strikes many of us today as plausible, it would not always have struck people that way. After all, one might act in lots of ways that don't make a significant difference to various serious moral problems but that nonetheless reveal that person's unreflectiveness, callousness, disrespect, or other kinds of viciousness. In Ancient Greece, for instance, these sorts of judgments—judgments of *character*—would have struck many as being morally primary, and so having just as much normative power as considerations of wrongness generally do.

The classical view underlying such a judgment is one that we can call an ethic of virtue, or simply 'virtue ethics'. It has quite the pedigree, going back at least to Socrates, Plato and Aristotle[1]; and although it has fallen out of mainstream favor somewhat, it has contemporary defenders as well.[2] According to a virtue ethical approach, focusing on individual *actions* is the wrong way to go about moral evaluation, as doing so leaves out much that is morally relevant. After all, the man who saves a drowning child for the sake of getting his name in the newspaper does the action that one is required to do—he saves the child—but he does not act *correctly*. And if this judgment seems intuitive, that is because we expect a person in this man's situation to act beneficently, or to save the child as a result of the character trait associated with helping others. According to virtue ethics, then, what one ought to do is secondary to *who one ought to be*; one ought to be a virtuous person, and then what one ought to do is simply whatever the person with the relevant virtues would do in such a situation. Thus, the man who rescues the child ought to perform the rescue from a stable character, and simply because he has the practical wisdom—what Aristotle called *phroenesis*—to see that rescue is what is called for in this situation.

There is a difference between Virtue Ethics—a moral theory that gives pride of place to virtue, like that espoused canonically by Aristotle—and the inclusion of virtue in one's moral analysis. While not many moral philosophers are Virtue

---

[1] See Plato's (2004) and Aristotle's (1999).

[2] For a contemporary revival of traditional virtue ethics, see Rosalind Hursthouse's (2002). As we will discuss below, there is also a prominent view of virtue that is specific to climate ethics, which is that of philosopher Dale Jamieson (2014).

Ethicists, few would deny that the language of virtue is relevant to our moral evaluation. And for present purposes, this more modest inclusion of virtue is all that's necessary. A question that seems morally relevant, then, is whether certain actions typically exhibit certain character traits, or whether an agent who has a particular virtue is likely to act in a particular way. And often, the answer is yes: a courageous person is one who would save the drowning child without hesitation, even at some risk, because the child is in danger. Could these sorts of considerations help us in our evaluation of climate or procreative ethics?

Philosopher Dale Jamieson argues that the answer is 'yes': that in the context of the massively complex and problematic issue of climate change, we in fact *must* revert back to the language of virtue in order to make sense of our individual moral burden. For it is true that something is morally amiss in the case where I gleefully and spitefully toss my recyclable materials into the trash can, even when it is right next to the recycle bin. But the problem is not that I harmed anyone by doing so, or even that I made a significant difference to the problem of waste management; the problem, Jamieson would say, is that such an act reveals my bad character, or my lack of what he calls 'green virtues'. By intentionally acting as I do, I respond to the wrong reasons, and fail to develop or reflect character traits such as humility, temperance, or mindfulness (Jamieson, 2014, p. 186).

This employment of specific 'green virtues', then, reflects a long and prestigious line of thinking that goes back at least to Ancient Greece, but is put to important new work in helping us to think about what goes wrong with an individual's moral life when she does not work towards conserving our planet. An interesting question for our purposes, then, is whether the green virtues have anything to say about procreation. Although Jamieson doesn't mention this topic, it seems plausible to me that there is a clear application.

Jamieson suggests that, because the problems of today are new, we will not only need classical virtues to fully explain the moral phenomena, but also virtues that have new content, as well as wholly new virtues. In these two categories, Jamieson offers the virtues of 'temperance' — an obviously traditional virtue, but which has new application in today's world — and 'mindfulness', which is a virtue specifically needed for the problems of our complex, global society. Both of these virtues seem likely to have some application to the context of procreation.

A temperate person seeks to live in moderation and to exercise restraint and good judgment. According to Jamieson, a temperate person living in our world would reduce her consumption and express careful consideration of her carbon footprint (Jamieson, 2014, p. 187). Perhaps the intemperance of certain purely luxurious activities (such as taking one's motorcycle to the track!) is what is wrong with them, despite them making no significant difference. But perhaps also, one can be a temperate person by simply not having very many such luxurious hobbies, and by avoiding ones that have very little hedonic payoff. Flying from Washington, DC to Paris, France just for the weekend, and just because one can afford it, would likely be very difficult for the temperate person to justify.

Mindfulness, unlike temperance, is not a rehabilitated traditional virtue, but a new one, necessitated by the complexity of our modern problems. In my example of spitefully refusing to recycle above, I made the case look especially problematic precisely because of the spite—I knew that I could recycle and that there were good reasons for it, but did the opposite for some very strange reason. But of course, most people don't spitefully refuse to recycle. They simply don't think about it. And they don't think about the cost (to the globe) of buying a less fuel-efficient car, or of taking excessive and unnecessary vacations. It simply doesn't occur to most of us that we are part of the problem *just by living* in the way that society has taught us. And so, Jamieson suggests that we need the virtue of mindfulness, which requires that a person "see herself as taking on the moral weight of production and disposal when she purchases an article of clothing (for example). She makes herself responsible for the cultivation of the cotton, the impacts of the dyeing process, the energy costs of the transport and so on" (Jamieson, 2014, p. 187). Such a virtue would have each of us visualize and accept the often 'invisible' costs of our actions to climate change, thereby requiring that we justify our lifestyles in a way that most of us do not.

Both of these candidate 'green virtues' push us in the general direction, I think, of small families. Mindfulness is exactly what is clearly needed for many people in the world regarding their procreative behaviors, since it is not recognized that the entire population bears the cost of each new person made. Each of us who has a child acts in a way that has a greater impact on the environment than any other action we take in our lives—indeed, more than all of our non-procreative actions combined—and yet we simply do not see this decision as being environmentally relevant, or as requiring a high justificatory burden. Mindfulness would require that this change, and that we each take on the responsibility of all future actions taken by our children, and our children's children. It is then difficult to see how someone could be mindful and yet have five children—or perhaps even three. After all, understanding the costs would require seeing that our population simply cannot grow while sustaining life for all others; indeed, our population cannot stay its current size while sustaining all others.

A temperate person, too, would likely have a small family, as the decision to have children would be subject to the same constraint and good judgment as any other costly action. A person who lives a 'green' lifestyle in all other aspects of her life, but then has many children, could not be seen as environmentally temperate, since the decision to have even a single child swamps all of the good that one can do through all of her other actions (or inactions). Temperance, in the era of climate change, would require that each of us include the environmental costs of child-bearing into our general lifetime calculus of resource use, and to justify having a child as part of an entire life that has a modest effect on the planet and others.

The mindful, temperate person, then, who both recognizes and accepts the justificatory burden of procreating, and who exercises restraint and moderation, would plausibly find it difficult to justify having more than one child. After all, once one has had the great fortune to start a family by having a child, adding another child remains exceedingly costly to the environment, but has much less value to the parents; while having a first child takes one from being non-parents to being parents, having a second child merely changes the size of one's family. Certainly, one could *want* more children, and could even believe that it is *valuable* for her child to have a sibling, but these sort of justifications fall far short of the kind of justification offered for having a first child; after all, having a first child fundamentally changes who one is, by changing her into a parent. Could, then, a person who is mindful of the costs of child-bearing, and who exercises constraint and good judgment really justify having another child after the first?

As in the previous chapter's discussion, I will not claim that green virtues obviously and demonstrably require that one have at most one child. Perhaps there is a case for having up to two children since, as we saw before, having two children commits one's family to not growing the population. Indeed, one of the benefits of a virtue theory is precisely that it does not provide us with hard and fast principles of the form 'Have no more than one child.' Perhaps there is a way to live a green, virtuous life by offsetting one's procreative behaviors in other aspects of one's life. What does seem clear, however, is that an individual with the green virtues would see procreation as a moral issue, and would take there to be value in having a 'small family', whatever precisely that turns out to mean.

5.2   Moral Reasons

In the previous chapter's discussion, I relied heavily on the concept of a 'reason' — a philosopher's term of art, to be sure, but also an intuitive way of talking about the considerations for and against particular actions. If acting within our rights is not 'right' in some particular case, then there must be something to be said against that action — that is, there must be good moral reason not to so act. And if it is sometimes *wrong* to act on one's rights, then this seems to have something to do with the reason on which one acted. A potential sexual partner's skin color is not a reason to refrain from sex with him, and so even though refraining from sex is within my rights, refraining for *that* reason seems wrong. What I want to point out in this final section is that reasons themselves make up an important aspect of the moral landscape, and that this is sometimes forgotten in discussion of obligation. That is: reasons for action *matter*, as they create a demand for *justification*. As a result, I contend, the arguments of this book constitute just such a demand in the case of procreation, and very many of us are not successfully responding. That is: very many of us are living unjustifiably with regards to our procreative lives.

The first thing to note in a discussion about our reasons is that we use the language of reasons in two, distinct, senses: there is a justificatory (or normative) sense, and there is a motivational sense. I will distinguish between them by calling the former *normative reasons*, and the latter *motivating reasons*. A normative reason, then, is what we tend to focus on in doing moral philosophy, as it serves as a justification. I have a reason not to buy an absurdly large home because of the effects that doing so has on the environment. Another way to put this claim is that the environmental effects of caring for a large home justify buying a reasonably-sized house. In the parlance of some contemporary philosophers, a reason is *a consideration that counts in favor of something*.

Normative reasons do not always motivate us to act, and sometimes we are motivated by considerations that aren't, we think, real justifications. But we often still call the consideration that motivated someone 'his reason'. In this case, we are talking about motivating reasons. So if Chris bought a large house in order to make his colleagues feel bad about their relative lack of success, then we might say that hurting his colleagues was his reason for the purchase. But importantly, when doing careful moral philosophy, we don't want to thereby ascribe any justificatory power to the goal of hurting his colleagues' feelings. Chris's motivating reason *does not* justify his action, and so his motivating reason is not a normative reason. Indeed, this separation of the actual reasons that bear on house-buying and the considerations that motivated Chris seems to be a big part of what is wrong with Chris's action: there are very good reasons to buy a reasonable-sized house, but he, instead, bought a very large house with the goal of making others feel bad.

Although I'm not sure it's ultimately true, we might think that Chris, and we, have a (moral) right to buy whatever kind of home we can afford and that is for sale. If so, then we have another example where we can see a serious moral problem with someone acting within his rights. And our new-found reasons terminology helps to explain such cases: acting on a moral right doesn't imply either (a) that there are good (normative) reasons for so acting; or (b) that one's (motivating) reason actually justifies the act. Going back to the case of the person who refuses sex for racist reasons, this seems to be the problem here as well: we stipulated that there were good reasons to have sex (everyone would enjoy it, it would result in some good, whatever other consideration you like), and that the person withholding sex did so only because the potential partner was Black. What is wrong, in this case, is that the relevant normative reasons all seem to favor having sex, whereas the woman's motivation is not a normative reason—that someone is Black is no justification for not having sex with him.

The goal of the present section, then, is the modest one of simply pointing out that *reasons matter*. It seems to matter greatly to our moral evaluations what reasons there are, and what reasons one acts on. Importantly, I take it that these two considerations of our reasons provide us with two ways to understand our individual moral burden, even in the absence of duty or obligation.

When reasons count in favor of some action—say, preventing suffering—then performing some other action requires justification. That's because the reasons to prevent suffering justify the prevention of suffering, and so if there are not competing considerations to be raised on behalf of another action, then one simply acted unjustifiably; that is, one acted in a way that the reasons, or justifications, didn't favor. For instance: while we do tend to think that because choking someone causes her pain, we have a good reason not to choke others, this reason might be outweighed by the fact that someone is drowning, and that choking her into submission is the only way to calm her sufficiently to get her out of the water. If I have choked a drowning person in order to save her, she might say upon waking, "You choked me!" And this seems to call out for justification. But when I respond by saying, "Well, you were drowning, and if I allowed you to stay conscious, you would have drowned both yourself and me in your panic," I have discharged my justificatory burden. There *is* a reason not to choke the drowning victim, since choking is a harm; but it is outweighed in this instance by the reason to save her life.

In the case of procreating, the arguments of this book seem to suggest that there are many, good reasons not to have children. Having a child contributes to the massive, systematic harm of climate change (more so than any other non-procreative act a typical person takes), depends on and reinforces injustice, and puts the created child in harms way. While I admitted that these considerations may not entail procreation-limiting obligations, they do seem to constitute reasons, and these reasons create a justificatory burden for anyone who wants to have a child.

In the best, most seemingly-justifiable cases of procreation, one might respond to such a burden by claiming that creating a family in this way is a radical and unique good, the lack of which would render one's life deeply unfulfilled. And the existence of this sort of reason *to procreate* does seem important. Although I'm not sure whether it truly answers the demand for justification, given the reasons against procreating, it is at least a plausible candidate consideration. So it looks to me like an open question about whether it is generally justifiable to have a child; not everyone will take it to be such a good, and so not everyone will have a strong reason in favor of doing so. Further, the strength of the reason in favor seems to depend on whether or not one already has children, and how many children one has. It may well be that having a child is sometimes justifiable and sometimes not—that is, it may be that at least sometimes, the reasons count in favor of not having a child.

Note, however, that we so far have discussed only *what (normative) reasons there are*, and not what (motivating) reasons people actually act on. I am suggesting that there is a real moral issue if the reasons against procreating swamp the reasons in favor of procreating in certain cases, but there might be even more of a moral problem with the reasons that people *in fact act on* when they procreate. If a couple does not, for instance, take having children to be such an intense good, but has several children just because they don't care enough to think about whether they should, this seems to worsen the moral situation. There are still very weighty moral reasons against procreating, but this couple appeals to no good reason in favor of procreating. Or it could be even worse, in that the couple might have a child for suspect reasons, such as 'giving one's parents a grandchild', or maybe even 'creating a future soldier for the state'.[3] In such cases, it seems plausible to me both that the couple's actions are unjustifiable (since the reasons against clearly outweigh the reasons for), and that the motivating reasons on which they act tell us important things about them—perhaps that they see children as a means to other ends, rather than as ends in themselves.

Of course, the reasons for which many people procreate are not nearly so clear. Many people procreate because it's what one does at a certain stage in life, and many women in patriarchal societies have no control over whether and how often they reproduce. Further, unplanned pregnancies are not undertaken for any reason at all, although they are continued (rather than terminated) for some set of reasons. However, most of the people who would be in a position to read this book will, if

---

[3] Christine Overall references these and other suspect reasons for procreation in her (2012).

they procreate, procreate for reasons; that is, it will be a conscious decision, because it is desired or taken to be a good. To those individuals—to *us* (as I take myself to be in this group)—my argument is that we should take seriously the strength of the reasons in favor of not procreating, and we should inquire persistently as to whether the actual reasons on which we are acting are good ones.

5.3   Meaning and Blame

The previous invocation of 'one's reasons', or the reasons on which one acted, has been helpfully discussed by philosopher T. M. Scanlon as determining the *meaning* of an act.[4] According to Scanlon, the permissibility of an act is, as we have discussed, a matter of what reasons and obligations there are; if I have a duty to recycle, and there are no countervailing reasons, then failing to recycle would be impermissible. Permissibility, then, has nothing to do with my own psychology, or what I'm intending to do when I'm recycling or not.

Such internal considerations do, however, play a role in determining another morally relevant feature—what Scanlon calls the 'meaning' of an act. In short, Scanlon holds that we often want to know more about an act than whether it was merely permissible; we want to know how we should judge, react, or alter our relationship with the actor in light of the action. And this, he argues, we do not get merely from judgments of permissibility. What we need to know is what the agent 'intended' in a narrow sense—that is, what the agent took to be the reason for acting. And when we have this information, we have a much richer picture of the morality of the act. If we know that the stranger saved the child from drowning in order to earn local fame, then we take his act to be permissible, yes (after all, saving the child was the right thing to do!), but also morally corrupt in some way. The actor's reasons reveal his view of other people and vulnerable children, and we may want to modify our interactions with him. Whereas if the actor saved the child out of a sense of duty, or a recognition of the suffering of others, then again, we will want to assign special meaning to the act—it was not only permissible, but *good*.

It's worth noting that this special goodness of an act done for the right reasons has a name in the history of philosophy: such an act is said by Kant to have 'true moral worth' (Kant, 2012). And indeed, Kant made quite a big deal out of actions that have true moral worth, such that many philosophers think of Kant as the moral philosopher concerned with *intention*. However, what Scanlon shows us is that intention may not be relevant to at least one important part of morality—the part dealing with permissibility—but that it might tell us much about what an act *means* to us, the moral community.

Acts can be not only permissible or not, then; they can also be good and bad in certain ways, as they can have good and bad meanings for the moral community. On Scanlon's view, we can carry the moral evaluation a step further with these new

---

[4] What follows is a summary of large sections of (Scanlon, 2008).

tools, and employ them in order to determine whether we should *blame* certain actors for what they've done. This is because, while praise and blame tend to go along with judgments of permissibility (we blame people for doing wrong), these evaluations can, in fact, come apart. We withhold blame from some wrong actions because of what the actions mean, but we also blame people for permissible actions because of what they mean. Scanlon's particular theory is distinctly relational, holding that the act of blaming is the changing of our relationship with someone in light of the judgment that her act warrants such a change; however, the most important part of the view for our purposes is simply that an action for some particular reason can convey a certain meaning to others, such that it warrants our blaming the actor. On a theory like Scanlon's, then, we are able to account for a much richer moral evaluation of particular actions, accounting for not only permissibility, but also the particular meaning of an act, and the justification for blaming the agent for that act.

We can now go back to the concluding judgment of the previous section, and ask what the reasons for which many of us procreate tell us about the morality of our procreative behavior. Even if we are within our rights to have a child, we noted that doing so may not be favored by the reasons, and that this by itself is important. However, we can now see also that procreating *for certain reasons* can be morally relevant, as it can determine the meaning of the act, thereby determining how the rest of us judge the procreator. If, as seems plausible, creating a child can be better or worse—if it can be done for better and worse reasons—then ideally we would want to procreate only for *good reasons*, and not simply because 'we have the right to do so'.

This is all perfectly vague advice, however, and so before concluding this final chapter, I will look at a few, fairly typical cases of procreative reasoning, and provide a brief analysis of each. My hope is that doing so will begin to make sense of what I want to call a 'moral burden' to have small families, which exists even if there is no strict duty to have a family of any particular size.

5.4 The Moral Burden to Have Small Families

It is surprisingly common, in my experience, to hear successful, educated people talk of their decision to have a child as bound up in expectation: it is time to do the next thing in our lives. Now, if this sense of expectation is combined with the intense excitement of being new parents, then it might not be worrisome; but sometimes it's not. Indeed, sometimes young couples feel pressured to have children by parents, society, one's 'biological clock', or the threat that they will come to regret a decision not to. Indeed, in twenty-first century America, it is becoming ever more common for women in particular to worry about what pregnancy, child-birth, and parenting will do to their promising careers, and for parents in general to worry about the disappearance of their social lives.

When a sense of expectation is enough to drive having a child—especially when the parents aren't even sure they want everything that goes along with parenting—

this seems like a morally worrying situation. We know by now that the reasons not to have a child are compelling, and the demand that they make for justification is not met by this lukewarm sense that having a child is probably 'the thing to do'. Taking such an expectation to be a reason to create another person doesn't seem to take seriously enough the nature of the act. We might worry that such a couple is insufficiently reflective and thoughtful about such a morally serious choice. This lack of moral seriousness may indicate a character flaw, and may tell the rest of us something about such a couple—about how importantly they take the interests of the population—and if the disregard is egregious enough, we may even be justified in modifying our relationship with them.

Of course, the more common case (I think, and would hope) is one in which prospective parents are thrilled at the prospect of fulfilling one of their lifelong goals. The pregnant woman is excited to experience gestation, and to see what it is like to create new life; and her partner is equally excited to be as active as possible. They both, then, can't wait to embark on the amazing journey of parenthood. It seems difficult to fault the prospective parents in such a case. They may even understand some of the reasons against having a child, but take it that such a unique, meaning-granting project should not be denied to anyone. Such a couple is standing up to the demand for justification, and providing a powerful case.

As we saw in evaluating the candidate obligations in Chap. 3, however, a really difficult question concerns whether anyone has good reason to have more than one child. So suppose that we have a morally sensitive, thoughtful couple like that above, and they enjoyed having the first child so much that they want to have more. Further, they believe that their first child will benefit from having siblings, and that their house will be more full of love and joy as more children are added. If the value and goodness of creating and raising a child constitutes a good reason in the first case, does it not constitute the same reason in every case?

Unfortunately, it seems to me that it does not. The more children that one has, the stronger the reasons become against having another. Having two children contributes more to climate change and injustice, risks harm coming to two new people rather than one, and becomes less temperate and mindful; further, having a first child makes a nearly unprecedented change in one's life, taking one from being a non-parent to being a parent. If becoming a parent is valuable, then having a first child has a huge value, in that it gives one this new, desired identity. But being a parent of two children, or three children, or many children, does not seem to have the same kind of unique value. While procreating may be the kind of project that meets steep justificatory demands, procreating a certain number of times may not.

I don't know whether this means that having exactly one child is uniquely justifiable. Perhaps a case can be made for two, arguing that siblinghood is valuable, and that having no more than two children still commits one to not growing the population. This still seems like a more difficult case to make. And I start to have real trouble coming up with plausible justifications after two. Some may argue for a particular value of 'big families', or they may argue on religious grounds. The stakes in this debate are so high that I have a hard time imagining that such arguments meet the justificatory demand made by compelling reasons not to procreate. This does not mean that there are no such arguments—perhaps my inability to see more of them is a failure of moral imagination. What it does seem to indicate, however, is that those arguments are rarer and more difficult than most of us would have imagined, as there is a weighty set of reasons that puts pressure on each of our decisions to procreate.

This general justificatory pressure is what I think of as the moral burden to have small families. There may be no obligation to have a family of some size, as we saw in our discussion of integrity and procreative rights. But we do seem to bear the burden of powerful moral reasons that count against procreation. And it's a burden to have 'small' families rather than to have no children, or only one child, because contexts differ, and so different families may plausibly be able to provide different justifications. For some people, having a child is difficult to justify; for others, having more than one or two children is hard to justify. While I have left it open whether there might be plausible justifications for having more than two children, I have suggested that the moral case against procreating gets stronger the more children one has.

5.5 Conclusion

There are too many people on Earth, with many of us using far too many resources. One precious resource that has been terribly depleted is the atmosphere's ability to absorb CO_2 without violently altering the global climate. As a result, we now face dangerous climate change, with catastrophic climate change on the horizon if we do not take decisive action. This, I have argued, makes overpopulation a public health crisis, in desperate need of addressing.

To many of us, sober facts about massive threats like climate change seem to entail individual moral obligations to contribute towards mitigation efforts. We might then expect that the massive threat of overpopulation entails an individual moral obligation to limit one's procreative actions. However, there is a powerful argument against such an intuition: namely, that because our individual actions cannot make a meaningful difference to a problem the scale of climate change or overpopulation, we cannot be obligated to refrain from particular actions.

I have argued that this objection fails. It fails because we do not *only* have duties related to goodness. We also have duties that do not depend on an act's consequences, and these plausibly include the duties to refrain from contributing to massive systematic harms, to refrain from committing injustice, and to protect our children. Unfortunately, all of these candidate duties seem to imply procreation-limiting obligations.

It is, however, worrisome to consider that procreation-limiting duties might threaten both one's integrity, and one's reproductive autonomy. Although I am unsure that such worries get us completely off the hook from procreative obligation, for the purposes of this project, I have conceded that we may not have such obligations. Is there nothing we can say, then, about the morality of procreation?

On the contrary, I think that even if we have no procreative obligations, it is very likely that we have a 'moral burden' regarding our procreative acts. Acting within one's rights does not mean that one is acting rightly, and so we must ask what could account for the sense of 'right action' that is divorced from mere permissibility. One way to account for such a burden would be to adopt the language of Green Virtue from Dale Jamieson, which has a lot of explanatory power concerning how certain procreative acts reflect on one's character.

Perhaps even more importantly is that the fact of procreative rights, and the corresponding lack of duty, does not entail that there aren't *very good moral reasons* not to act on one's rights. The arguments in favor of procreation-limiting moral principles—even if they do not generate individual obligations—seem to generate very compelling reasons to limit one's procreative behaviors. Further, then, many people seem not to have good reason to procreate at all, and the availability of such reasons seems to diminish as one has more children. When an individual or a couple takes a weak consideration to be a reason for doing something as morally serious as creating new human life, this act *means* something to the rest of us, as the individual or couple has made it clear where the interests of the moral community stand in their evaluation. In egregious cases, this may even justify the rest of us blaming actors who procreate badly (even if permissibly).

We are left, I think, with a moral burden to have small families. The powerful reasons in favor of limiting procreation generate a demand for justification; if one fails to meet this demand, then her procreative activity is morally unjustifiable. And meeting this demand, I think, becomes progressively more difficult as one has more children. Given the moral burden to have small families, having any children at all may well be unjustifiable for many people; for some of the rest of us, the case for having one child seems fairly compelling. Might some people be justified in having more than one? Perhaps. But the burden is on them to make the case.

Morality has more in its arsenal than merely obligation, duty and rights; reasons can burden us, and acting justifiably looks, to me, to pressure us towards small families.

References

Aristotle. (1999). *Nichomachean ethics: With introduction, notes, and glossary* (2nd ed.), (T. Irwin, Trans.). Indianapolis, IN: Hackett

Hursthouse, R. (2002). *On virtue ethics*. Oxford: Oxford University Press.

Jamieson, D. (2014). *Reason in a dark time: Why the struggle against climate change failed—and what it means for our future*. Oxford: Oxford University Press.

Kant, I. (2012). In M. Gregor, & J. Timmermann (Eds.), *Groundwork of the metaphysics of morals* (2nd ed.). Cambridge: Cambridge University Press.

Overall, C. (2012). *Why have children? The ethical debate*. Cambridge: MIT Press.

Plato. (2004). *The republic* (C. Reeve, Trans.). Indianapolis, IN: Hackett

Scanlon, T. M. (2008). *Moral dimensions: Permissibility, meaning, and blame*. Cambridge, MA: Harvard University Press.

Index

T.N. Rieder, *Toward a Small Family Ethic*, SpringerBriefs in Public Health,
DOI 10.1007/978-3-319-33871-2

Printed in the United States
By Bookmasters